Leibniz and the Environment

The work of seventeenth-century polymath Gottfried Wilhelm Leibniz has proved inspirational to philosophers and scientists alike. In this thought-provoking book, Pauline Phemister explores the ecological potential of Leibniz's dynamic, pluralist, panpsychist metaphysical system. She argues that Leibniz's philosophy has a renewed relevance in the twenty-first century, particularly in relation to the environmental changes and crises that threaten human and nonhuman life on earth.

Drawing on Leibniz's theory of soul-like, interconnected metaphysical entities he termed 'monads', Phemister explains how an individual's true good is inextricably linked to the good of all. Phemister also finds in Leibniz's works the rudiments of a theory of empathy and strategies for strengthening human feelings of compassion towards all living things.

Leibniz and the Environment is essential reading for historians of philosophy and environmental philosophers, and will also be of interest to anyone seeking a metaphysical perspective from which to pursue environmental action and policy.

Pauline Phemister is Reader in Philosophy at the University of Edinburgh, UK. She is author of *Leibniz and the Natural World* and *The Rationalists: Descartes, Spinoza and Leibniz*. Among her other books is a coedited interdisciplinary volume *Human–Environment Relations: Transformative Values in Theory and Practice*.

Leibniz and the Environment

Pauline Phemister

 Routledge
Taylor & Francis Group

LONDON AND NEW YORK

First published 2016 by Routledge

2 Park Square, Milton Park, Abingdon, Oxfordshire OX14 4RN
711 Third Avenue, New York, NY 10017

Routledge is an imprint of the Taylor & Francis Group, an informa business

First issued in paperback 2018

British Library Cataloguing in Publication Data
A catalogue record for this book is available from the British Library

Library of Congress Cataloging in Publication Data
Names: Phemister, Pauline.
Title: Leibniz and the environment / Pauline Phemister.
Description: 1 [edition]. | New York : Routledge, 2016. | Includes
bibliographical references and index.
Identifiers: LCCN 2015036213| ISBN 9781138924918 (hardback : alk. paper) |
ISBN 9781315684086 (e-book)
Subjects: LCSH: Leibniz, Gottfried Wilhelm, Freiherr von, 1646-1716. |
Philosophy of nature. | Ecology.
Classification: LCC B2598 .P54 2005 | DDC 193--dc23
LC record available at http://lccn.loc.gov/2015036213

ISBN: 978-1-138-92491-8 (hbk)
ISBN: 978-1-138-58082-4 (pbk)

Typeset in Sabon
by Taylor & Francis Books

In loving memory of my father
Bruce Maxwell Hill Walker
1920–2012

Contents

Acknowledgements

The idea behind this book dates back to the early 1990s, to a time when I was just beginning to question my previously unquestioned assumption that Leibniz's philosophy is an idealist philosophy in which the physical world is merely a phenomenon or appearance of an underlying psychical reality. As I reinterpreted Leibniz's writings along realist lines, taking seriously his insistence that every being perceives the world through, or from the perspective of, its own body, I became increasingly conscious of the implications of Leibnizian realism for ecological philosophy. Arne Naess had drawn on Spinoza's philosophy in this respect and I began thinking about how Leibniz's thought would fare in a similar project.

However, it was only through various interdisciplinary environment research projects and reading groups at Edinburgh University's Institute for Advanced Studies in the Humanities, to which I was myself seconded from my department as Deputy Director in 2009, that my ideas began properly to take shape. I am deeply indebted to the Institute staff, the director, the late Professor Susan Manning, the administrator Anthea Taylor, and the secretary Donald Ferguson for creating the welcoming, relaxed environment in which through friendly and stimulating exchanges new ideas are generated and grow. I am equally indebted to all the colleague-friends who participated in – and in some cases organized – the aforesaid Institute activities, in particular to Emily Brady, Tom Bristow, Sarah Buie, Jonathan Delafield-Butt, Jeremy Dunham, Timothy Engström, Tim and Reiko Goto-Collins, Jane Goldman, Rachel Harkness, Wallace Heim, John Llewelyn, Leemon McHenry, Michela Massimi, Anna Ortin, Andrew Warren, Françoise Wemelsfelder, Wendy Wheeler and Catherine Wilson.

I am immensely grateful to the Leverhulme Trust for the award of a Leverhulme Research Fellowship in 2012. This freed me from administrative and teaching commitments and enabled me to bring my fledgling ideas together in a draft manuscript. Revision of chapter 8 was assisted by research funding I received as Co-investigator on the AHRC (Arts and Humanities Research Council) project, Caring for the Future through Ancestral Time. I am grateful to the AHRC and to my fellow investigators for making this project possible and for making it overflow with exciting and novel ideas.

With kind permission from the publishers, I have made use of excerpts from *Human–Environment Relations: Transformative Values in Theory and Practice*, chapter 2, 'Relational Space and Places of Value', 2012, pp. 17–30, Emily Brady and Pauline Phemister (eds), © Springer Science+Business Media B.V. 2012.

I would like to take this opportunity to thank my editors at Routledge, Tony Bruce and Adam Johnson, and my earlier editor, Tristan Palmer at Acumen. Without their assistance and encouragement this book would have been much longer in the making than it has already been. I am grateful also for the helpful comments of an anonymous referee.

I owe special debts to David and Maureen Cockburn and to Lloyd Strickland who read and provided greatly appreciated critical comments on the draft manuscript. I thank them too for their encouragement to see this project through to completion, and apologize for those places where I have not satis-factorily addressed their concerns. I give grateful thanks also to Michael McGhee for his generous help just when it was most needed.

By far my greatest debts are to my late father who passed his love of God and nature to his children and to my mother, Madeline, who replicated nature's beauty in our home. Thank you.

Abbreviations

A	Deutsche Akademie der Wissenschaften (ed.), *Leibniz: Sämtliche Schriften und Briefe*, ser. 1–7, Berlin: Akademie Verlag, 1923–. Cited by series, volume and page numbers.
AG	Ariew, Roger and Garber, Daniel (ed. and trans.), *Leibniz: Philosophical Essays*, Indianapolis and Cambridge: Hackett Publishing Co., 1989.
AK	Preußischen Akademie de Wissenschaften (ed.), *Immanuel Kant's gesammelte Schriften*, 29 vols, Berlin: Akademie Verlag, 1900–.
AT	Adam, Charles and Tannery, Paul (eds), *Oeuvres de Descartes*, 12 vols, Paris: Vrin, 1964–76.
BM	Baumgarten, Alexander, *Metaphysics: A Critical Translation with Kant's Elucidations, Selected Notes and Related Materials*, ed. and trans. Courtney D. Fugate and John Hymers, London: Bloomsbury Academic, 2013.
C	Couturat, Louis (ed.), *Opuscules et fragments inédits de Leibniz: Extraits des manuscrits de la Bibliothèque royale de Hanovre*, Paris: Alcan, 1903.
CPJ	Kant, Immanuel, *Critique of the Power of Judgment*, trans. Paul Guyer and Eric Matthews, Cambridge: Cambridge University Press, 2001.
CSM	Cottingham, John, Stoothoff, Robert and Murdoch, Dugald (ed. and trans.), with Anthony Kenny (vol. 3), *The Philosophical Writings of Descartes*, 3 vols, Cambridge: Cambridge University Press, 1985–91.
Curley	Curley, Edwin (ed. and trans.), *The Collected Works of Spinoza*, vol. 1, Princeton: Princeton University Press, 1985.
Dascal	Dascal, Marcelo (ed. and trans.), with Quintín Racionero and Adelino Cardoso, *The Art of Controversies*, Dordrecht: Springer, 2008.
DSR	Leibniz, Gottfried Wilhelm, *De summa rerum: Metaphysical Papers, 1675–1676*, ed. and trans. George Henry Radcliffe Parkinson, New Haven: Yale University Press.

Dutens	Dutens, Ludovici (ed.), *G. G. Leibnitii … Opera omnia: Nunc primum collecta, in classes distributa, praefationibus et indicibus exornata*, 6 vols (in 7), Geneva: apud fratres de Tournes, 1768.
Essay	Locke, John, *An Essay concerning Human Understanding*, ed. Peter H. Nidditch, Oxford: Clarendon Press, 1975. Cited by book, chapter and section numbers.
FC	Foucher de Careil, Alexandre (ed.), *Nouvelles lettres et opuscules inédits de Leibniz*, Paris: Durand, 1857.
Gebhardt	Gebhardt, Carl (ed.), *Spinoza Opera: Im Auftrag der Heidelberger Akademie der Wissenschaften*, 4 vols, Heidelberg: Carl Winters, 1925.
GM	Gerhardt, Carl Immanuel (ed.) *Leibnizens mathematische Schriften*, 7 vols, Berlin: A. Asher; Halle: H. W. Schmidt, 1849–63.
GP	Gerhardt, Carl Immanuel (ed.), *Die philosophischen Schriften von Gottfried Wilhelm Leibniz*, 7 vols, Berlin: Weidman, 1875–90; repr. Hildesheim: Olms, 1965.
Grua	Grua, Gaston (ed.), *G. W. Leibniz: Textes inédits*, 2 vols, Paris: Presses Universitaires de France, 1948.
H	Huggard, E. M. (trans.), *Leibniz: Theodicy; Essays on the Goodness of God, the Freedom of Man, and the Origin of Evil*, La Salle, IL: Open Court, 1985.
Klopp	Klopp, Onno (ed.), *Die Werke von Leibniz gemäss seinem hanschriftlichen Nachlasse in der Königlichen Bibliothek zu Hannover*, 11 vols, Hannover: Klindworth, 1864–84.
L	Loemker, Leroy E. (ed. and trans.), *Leibniz: Philosophical Papers and Letters*, 2nd edn, Dordrecht: D. Reidel Publishing Company, 1969.
LA	*The Leibniz–Arnauld Correspondence*, ed. and trans. Haydn Trevor Mason, Manchester: Manchester University Press, 1967.
LB	*The Leibniz–Des Bosses Correspondence*, ed. and trans. Brandon C. Look and Donald Rutherford, New Haven and London: Yale University Press, 2007.
LC	*The Leibniz–Clarke Correspondence*, trans. Henry Gavin Alexander, Manchester: Manchester University Press, 1956.
LO	Lennon, Thomas M. and Olscamp, Paul J. (ed. and trans.), *Malebranche: The Search after Truth; With Elucidations of The Search after Truth*, Cambridge: Cambridge University Press, 1997.
LS	*Leibniz and the Two Sophies: The Philosophical Correspondence*, ed. and trans. Lloyd Strickland, Toronto: Centre for Reformation and Renaissance Studies, 2011.
LV	*The Leibniz–De Volder Correspondence*, ed. and trans. Paul Lodge, New Haven and London: Yale University Press, 2013.
M	*Leibniz's 'Monadology': A New Translation and Guide*, ed. Lloyd Strickland, Edinburgh: Edinburgh University Press, 2014.

PW   *Leibniz: Philosophical Writings*, ed. G. H. R. Parkinson, trans.
            Mary Morris and George Henry Radcliffe Parkinson, rev. edn,
            London: J. M. Dent, 1973.
RB   Remnant, Peter and Bennett, Jonathan (ed. and trans.), *Leibniz:*
            New Essays on Human Understanding. Cambridge: Cambridge
            University Press, 1981.
SL   *Spinoza: The Letters*, ed. and trans. Samuel Shirley, Indianapolis:
            Hackett.
ST   *The Shorter Leibniz Texts: A Collection of New Translations*,
            ed. and trans. Lloyd Strickland, London: Continuum, 2006.
Treatise   Hume, David, *A Treatise of Human Nature*, ed. David Fate
            Norton and Mary J. Norton, Oxford: Oxford University Press,
            2000. Cited by book, part and section numbers.
WF   Woolhouse, Roger S. and Francks, Richard (ed. and trans.),
            Leibniz's 'New System' and Associated Contemporary Texts,
            Oxford: Clarendon Press, 1997.

Introduction

My aim is not to fill shops with futile books, written in the air, but where possible to provide something of use.

(Leibniz to Duke John Ferdinand, November 1671: Jordan 1927, 38)

As for me, I lay little importance in refutation but much in exposition. When a new book reaches me, I search for what I can learn, not for what I can criticize in it.

(Leibniz to Gabriel Wagner on the value of logic, 1696: L 470)

The incredible, and sometimes incredulous, system of philosophy of Gottfried Wilhelm Leibniz (1646–1716) draws eclectically from ancient and mediæval texts of the Western tradition, from Chinese and Arabic writings, as well from the more modern philosophical and scientific tracts of his own times and stands as a magnificent testimony to Leibniz's policy, described in the quotations above, of prioritizing the search for truth over criticism of others' errors. But not even in his most wildly optimistic moments could Leibniz have imagined the full extent to which his idealistic youthful aim to bequeath to the world 'something of use' would be realized after his death.[1] Yet, from the eighteenth century on, Leibniz's thought has proved inspirational to thinkers and practitioners, to philosophers and scientists alike. Approaching his extant corpus in the same spirit of openness and genuine seeking of truth as he himself employed in his reading of others, readers of Leibniz as diverse as Christian Wolff and Gilles Deleuze, Johann Bernoulli and Benoît Mandelbrot, have built upon his ideas, taking them forward in novel ways and ensuring that Leibniz's labours were not in vain.[2]

This book is predicated on the belief that Leibniz still has a great deal to offer in the twenty-first century, particularly in relation to the environmental crisis that now threatens human and nonhuman life on earth. His philosophy provides a set of metaphysical beliefs about the fundamental nature of reality, an integrated system of ethical and aesthetic values, and a vision of the world's perfection and the inherent goodness of all its constituent parts. Together, they provide a theoretical structure and a practical system of values to assist us as we attempt to repair humanity's fractured relationship with the natural world and instigate changes to the patterns of human behaviour that have played no small part in bringing us to this critical juncture.

An illustration from the philosophy of science can help us understand the necessity for deep changes at this time. In the middle of the last century, Thomas Kuhn described the history of modern science as extended periods of 'normal science' intercepted by 'scientific revolutions' in which the prevailing normal scientific paradigm is overthrown, the tipping point into revolutionary science occurring when the anomalies and counterexamples to the normal scientific paradigm become so numerous and so internally intractable and destructive that they can no longer be either ignored or accommodated within the standard paradigm, forcing instead the development of a radically new paradigm and the completion of the transition to the next period of 'normal science' (Kuhn 1970 [1962]). Kuhn conceived his account as a description of research methodology in the natural sciences. However, his theory has wider applicability. Just as the accumulation of anomalies and contradictions in the physical sciences periodically necessitates the construction of new overarching paradigms within which the natural sciences can flourish once again, so too the accumulation of individual environmental crises has now reached such dramatic proportions and the ruptures in our relationship with the living and nonliving world are now so great that new philosophical paradigms must be explored to replace the current paradigm within which these crises and ruptures have arisen. Anthropogenic climate change, pollution to land, air, water, space debris, ozone depletion, light and noise pollution, species depletions and extinctions and consequent biodiversity loss, deforestation, land erosion and drought are only some of the features of the environmental crises that bring in their wake a host of social, economic and political problems such as poverty, famine, disease, migration and war, and that threaten the very sustainability of life as we know it on earth. In the face of such extreme environmental disruption, it is imperative that we re-envision our human selves and nonhuman others as individuals in relation to each other within the wider whole and construct new paradigmatic frameworks that assist and encourage us towards behaviours that are at best positively life-affirming or at least 'do no harm'. What is needed is a revolutionary re-evaluation of the human–nature relationship and the construction of a new metaphysical-ethical-aesthetic paradigm that generates and supports human ways of living that harmonize with and promote the flourishing of human and nonhuman life and environments. In essence, this demands a sound ecological philosophy – a philosophy that, as the derivation of its name suggests,[3] focuses on the symbiotic relationships of living organisms and their environments.

The creative construction of alternative ecophilosophical world-views belongs to deep, as opposed to shallow, ecology. The distinction was first articulated by the Norwegian environmental philosopher and activist, Arne Naess (Naess 1973). Shallow ecology is characterized by its tendency to focus on finding solutions to immediately pressing environmental problems, such as the development and promotion of alternative wave, wind and solar energy production over carbon-based coal and oil. Although vital, Naess believed that the shallow ecological approach, in seeking to address one-by-one the causes of particular environmental issues, fails to tackle the general cause at the root of all the particular manifestations: our failure to critically examine, evaluate and revise our fundamental metaphysical assumptions about the nature of the world and our place, role and value within it. The deep ecological approach presupposes a close connection between human choice and action and the fundamental beliefs or assumptions that constitute the framework within which these choices and actions are realized. In this spirit, deep ecologists maintain that nothing less than a radical overhaul of the most deeply held metaphysical convictions that underpin our conscious choices and behaviours can produce the sustained long-term changes to our lifestyles that are needed if we are to adapt to our changing environment and prevent or hinder further degeneration.

The deep ecological philosophy – or personalized 'ecosophy'[4] – that results from this undertaking must give metaphysical grounding to key environmental and ethical concepts such as interconnectedness, bioegalitarianism, and diversity. To this end, deep ecological philosophers have typically turned for inspiration to the grand speculative metaphysical system of the seventeenth-century Dutch thinker, Benedict de Spinoza.[5] Undoubtedly, Spinoza's absolute monism with its identification of God as Nature from whose essence flow the myriad living things that comprise the infinitely extended, thinking organism that is our world provides a powerful sense of the interconnectedness of all things, of the intrinsic value of each, and the important and unique contribution each living thing makes to the dynamic, organic whole. But although Spinoza's monism undoubtedly has tremendous ecophilosophical strengths, its sublimation of the individual within the whole worryingly and unhelpfully undermines both the individuality and the agency of finite things in the Spinozan world. Relatedly, the absolute necessitarianism of Spinoza's system under which every single event that occurs follows directly from the divine nature and could not have been different in even the tiniest detail poses a serious threat to the concept of the human individual as a free agent. All that happens, all of the apparent choices we make, however reasonable, are ultimately determined by the nature of the larger world-organism. Such strict necessitarianism also casts a shadow over the value and usefulness of imagining different futures as we try to envisage new ways of being.[6]

When seeking alternative pluralist metaphysical systems in the history of philosophy, philosophers have turned, not to the dynamic pluralist system of Spinoza's contemporary, Leibniz, but to the speculative and panpsychist process philosophy of the early twentieth-century British philosopher, Alfred North Whitehead.[7] Certainly, as Susan Armstrong-Buck has shown, Whitehead's philosophy gives more than adequate support to notions of interconnectedness and the intrinsic value of all living things. Meanwhile, his idea of nature as dynamic and his conception of the ongoing interplay of 'actual occasions' that possess both mental and physical features that creatively generates new events in a universe whose future is open-ended are ecologically attractive features of his system that have the potential to avoid the necessitarian problems associated with Spinoza's monism. Nevertheless, like that of Spinoza, Whitehead's philosophy ultimately lacks a robust sense of the individuality and independent agency of living beings. Whitehead rejected early modern notions of enduring substances. Instead, he proposed a pluralism of events and of fleeting, momentary actual occasions or experiencing centres of force and offered a reductionist account of the formation of individuals as arising out of the coming together of actual occasions.[8] However, this implies that an individual organism is dependent entirely on the elementary actual occasions of which it is composed. In this sense, the parts would seem to determine the whole rather than the whole determining the parts. This in turn makes it difficult to give a plausible account of how the entity as a whole being can be in control of itself or of its own actions. Whitehead attempted to address this by postulating that each individual being, constituted by actual occasions, possesses an overarching 'subjective aim',[9] but the threat to the individual's own freedom and agency remains, for an organism's subjective aim must ultimately result either from its aggregate parts or from God.

Although the Cambridge philosopher owed an immense intellectual debt to his German predecessor,[10] environmental and deep ecological philosophers have, by and large, passed over Leibniz. This is both unjustified and lamentable, not least because, as we shall discover, the Leibnizian metaphysics retains advantages of the Spinozist and Whiteheadian systems – such as the focus on activity and interconnectedness – while also accommodating a strong sense of the agency of individuals. Moreover, Leibniz's philosophy incorporates a teleological dimension that Spinoza's lacks and includes an aesthetics that is absent from the systems of both Spinoza and Whitehead.

The neglect of Leibniz has not been total. An early study by Walter O'Briant (O'Briant 1980) surveyed various environmentally helpful aspects of Leibniz's thought, paying especial attention to Leibniz's appreciation of the interconnectedness of all things and his conception of the perfection of the world measured in terms of its order and variety, a notion that bears directly on the value of biodiversity and which we shall find is also hugely significant with respect to the objective aesthetic value of living and nonliving things. The publication of O'Briant's article coincided with the appearance of Carolyn Merchant's ground-breaking critique in *The Death of Nature* (Merchant 1989 [1980]). Merchant highlights the importance of a Leibnizian vitalism in support of the environmental cause. To the aspects identified by O'Briant and Merchant may be added such features as Leibniz's pluralism, his defence of individual freedom and responsibility, his recognition that all living things are embodied, his theories of love and justice (suitably adjusted), his account of beauty, his belief that the world operates teleologically as well as mechanically, and his overall optimism. Through chapters 4 to 8, we shall discover their importance and usefulness for the development of a Leibnizian ecological philosophy.

For the most part, however, the reception of Leibniz in the environmental community has not been favourable. Deep ecologist George Sessions has accused Leibniz of an anthropocentric instrumentalism regarding the natural world.[11] O'Briant too questioned the ecological credentials of some of Leibniz's views, in particular his denial of causal interaction between individuals

(O'Briant 1980, 220). Leibniz claimed that the metaphysical unities that comprise our world – the monads – 'have no windows through which something can enter or leave' (*Monadology* §7: GP VI 607; AG 214), from which it follows that there is no causal interaction between monads and that each is radically independent of the others, existing as if it were 'a world apart' (*Discourse on Metaphysics* §14: GP IV 439; AG 47). Other aspects of Leibniz's thought are equally problematic. For instance, while the 'harmony' in his trademark doctrine of pre-established harmony is laudable, the 'pre-established' aspect presents a challenge insofar as it maintains that the entire history of the world is already known and approved by God and was so even before anything had been created. But if the future is already certain and cannot be other than it will be, will all our attempts to alter its current direction of travel ultimately prove pointless? Leibniz's doctrine that of all possible worlds, the existing world is the best possible raises the same dilemma in a slightly different way: if this world is the best possible, there would seem to be no ground for concern over the damage we presently inflict.

Perhaps, however, the anti-ecological assessments of Leibniz's thought are overstated. By approaching Leibniz's own works in the manner he employed in his reading of others, seeking not so much to criticize but to learn, we will discover that a more nuanced interpretation of the Hanoverian elicits a great deal that is helpful in the construction of a new ecometaphysical perspective. Thus, in chapter 4 Sessions' charge of Leibnizian anthropocentrism is mitigated, albeit only completely removed with the adjustment to Leibniz's views on love and universal justice that are introduced in chapter 6. In chapter 5, O'Briant's concern about the lack of causal interaction is rendered harmless by Leibniz's relational account of individual identity which shows that, far from being independent of each other, the very essences of living things are so intimately bound up with each other that they may be said to be inherently and unavoidably codependent. For Leibniz, every single living thing is a mirror in which the whole universe is reflected. Leibniz's monads may not have windows, but their representative natures ensure that at every moment, each constitutes a unique expression of the present state of the entire universe. The issue of pre-established harmony and the certainty about the future that it entails is addressed in chapter 8, where it is argued that the doctrine is compatible with human freedom and the continuation of our efforts to ensure the best possible future outcomes for all.

We begin, however, with Descartes and Spinoza (chapter 1). Recent analyses have shown that Descartes is not nearly as villainous in environmental terms as he has been painted, but still there is enough truth in the standard story to explain his generally less than favourable reception in environmental circles. Chapter 1 outlines the central points of concern. However, the main focus in this chapter is on Spinoza and the adoption of his monist metaphysics by deep ecologists. Following a brief outline of Spinoza's philosophy and an account of the principles or 'platform' of the Deep Ecology Movement (DEM) founded by Naess and Sessions, our attention turns to Naess's own Spinozist ecosophy

(Ecosophy T) and an assessment of its success in providing a sure foundation, as Naess proposes all ecosophies should, for the DEM platform.

Chapter 2 opens the discussion of Leibniz with an overview of what are for our purposes the most salient features of his philosophy, including his concept of the 'monad' as a 'metaphysical atom' or indivisible unity, a perceiving, appetitive centre of force endowed with an organic body. There follows in chapter 3 an indicative, but unavoidably incomplete, account of ways in which Leibniz's ideas helped shape philosophical thought in Germany, France and Britain through the eighteenth to the twentieth centuries. Of the many philosophers who could have been discussed in this chapter, attention has been focused on those for whom Leibniz was a predominantly positive influence and who developed, amended and revised his ideas especially in respect of those topics, such as plurality, activity, perfection, harmony and progress, that feature significantly through the rest of this volume. Even so, some will still be absent whose inclusion might have been justified. Nevertheless, I hope that the reader comes away with a general sense of the variety and extent of the influence of Leibniz's thought on issues connected with the themes we explore in the chapters that follow.

Chapter 4 sees the start of our closer examination and exploration of ecophilosophically significant features of Leibniz's thought. In this chapter, we investigate the distinction between organic and inorganic bodies and find in Leibniz's vision of a world comprising an infinite plurality of living ensouled individuals grounds for a bioegalitarianism whereby all living creatures might in principle be accorded equal weight and consideration. Building upon this, and drawing on Jane Bennett's notion of vibrant matter, I propose a further notion of 'ontoegalitarianism' whereby inanimate things – which for Leibniz are aggregates of ensouled (i.e. animated) living things and which are, moreover, integral parts of the environments within which all living things are intrinsically embedded – might also be included in principle within the sphere of our care and consideration. The intrinsic relationality of the natures of Leibniz's individuals is taken up in chapter 5. Here, in a departure from the traditional understanding of intrinsic properties as nonrelational, it is explained how Leibniz envisaged the monads' perceptions and appetitions as qualities that are both intrinsic or internal and relational. In light of this, I outline an account of the value of each Leibnizian individual that overturns the common understanding of intrinsic and instrumental values as mutually exclusive. Instead, intrinsicality and instrumentality are combined within a notion of 'relational value', according to which the instrumental value that each individual has in relation to others and to the world as a whole forms the basis for its (relational) intrinsic value. Chapters 6 and 8 develop these ideas of relationality and value in the context of Leibniz's theories of relational space and time respectively. Leibniz conceived space as arising from the relations that hold between coexisting bodies. In chapter 6, I use Leibniz's theory of physical, extended space as a model to develop a theory of space as having psychical as well as physical dimensions, with the psychical dimensions arising from the relations that hold between coexisting perceiving and appetitive living beings. In consequence, it becomes possible to describe and evaluate space and particular places not only quantitatively in terms of measurable areas, but also qualitatively in terms of values such as justice, love, charity or universal benevolence. Chapter 7 concerns the expressive or mirroring relations that hold between minds or souls, between bodies, and between souls and their bodies in the Leibnizian world. Understanding these, with Leibniz, as forms of nonverbal communication (communication-as-harmony), a theory of empathy is outlined. In this, it is proposed that our expressive relationships with others enable us to, as it were, enter the 'place of the other', the external appearances of their bodies acting as generally reliable indicators of their inner experiential[12] lives. In this chapter, empathic communication is explained in respect of empathy with others existing at the same time, but in chapter 8, the possibility of developing empathic relations with past and future beings is raised. The main concern of this final chapter, however, is Leibniz's account of relational time. As in the discussion of space, my remarks on time focus on the qualitative aspects of relations between living beings, namely, the perceptual and appetitive relations that link the present to the past and to the future. The axiological dimensions of time, the influence of the past on the present and the shaping of the future in the present naturally bring into focus the question whether individuals and the world as a whole progress over time. Leibniz's optimistic view is that although in one sense the degree of perfection of the world remains the same at each moment, in another sense, building upon past experiences, individuals may be said to become increasingly more perfect as time moves on and the world itself may be regarded as increasing in perfection as it unfolds. Moreover, although the future is already certain, how it unfolds depends upon our and others' actions in the present. I bring this chapter and the book to a close by proposing that we look to the practice of gardening as our guide to living well in the present and optimizing probabilities of good outcomes in the future.

On occasion, I revise or develop certain of Leibniz's opinions. I have already mentioned one instance: in chapter 6, I expand Leibniz's theory of physical space to incorporate the idea that space and places are qualitatively rich. Another example found in this same chapter is the revision of Leibniz's conception of universal benevolence to apply not only to rational beings, but to all living things, while in chapter 8, I employ the injunction, voiced by the character Candide in Voltaire's satire of Leibnizian optimism, to 'tend the garden' in order to elaborate a conception of the best possible world as a world that needs to be attentively tended and lovingly nurtured. At other times, I recast Leibniz's views in a different light, as for instance in chapter 5 where I use his account of individuals' relational natures in the development of a theory of relational value and in chapter 7 when I consider his proposal that we put ourselves in the place of the other in the context of a theory of compassionate empathy. Leibniz himself, we may presume, would not have objected to the adaptation of his views. It is, after all, as we saw at the beginning of this introduction, a procedure he himself favoured and one that is in keeping with his expressed desire that his writings be put to good use.

Nevertheless, I have retained Leibniz's Judaeo-Christian understanding of God as the omnipotent, omniscient and all-benevolent creator of the universe, and generally endorse the fundamental role that God plays in his metaphysical system. To some secular readers, this may make Leibniz's philosophy completely unavailable. But it would be a great pity to disengage from the content of Leibniz's thought because it is presented in a seventeenth-century theological frame. After all, Leibniz's God is essentially a rational God. God's reason is founded on the Principles of Contradiction and of Sufficient Reason, but these '*two great principles*' form the basis for our reasoning too (*Monadology* §31: GP VI 612; AG 217). They are principles of human reason as much as they are of divine reason. By the Principle of Contradiction we know necessary truths, such as those of mathematics. However, the reasons for contingent truths, for things and events that could have been different, are sufficient reasons. These are the kinds of reasons we give when we try to explain why we acted as we did when it was entirely possible we could have acted otherwise. We say, for instance, that we acted as we did out of love or because it was the wise and just thing to do. We explain that we like a particular thing because it is beautiful or that we desire a certain outcome because it will increase diversity, restore order or bring harmony. We defend a decision by saying it was the best of the available options. It is precisely these concepts – of beauty, perfection, order, harmony, wisdom, love and justice – that Leibniz employs in describing his vision of the best possible world and in his account of how we must act if this vision is to be fulfilled. On account of this and his equally insightful views of the nature of life as active force, the intrinsically relational identity of individuals, the distinction between the natural and the artificial, and so forth, Leibniz's philosophy has value and interest as much for the secular environmentalist as it does for the theist.

My intention in this book has been to explore the ecological potential of Leibniz's dynamic, pluralist, panpsychist metaphysical system. I have refrained from entering into specific debates within the huge body of environmental literature. Nor have I entered into debates within Leibniz scholarship on specific points of interpretation of Leibniz's texts. Thus, while I support my own reading of Leibniz with reference to his texts, I do not here defend it against contrary interpretations.[13] I have paid special attention to Leibniz's views on the activity, natural embodiment and relationality of all living creatures, his notions of perfection, harmony, beauty and progress, as well as to the role of final causation or teleology in his philosophy. I hope to have sown seeds that will encourage others to explore further the ecological potential of Leibniz's thought and to advance it in intriguingly new directions as we reassess ourselves in relation to each other within this constantly changing world.

Notes

1 For a moving and informed account of Leibniz's final years, see Antognazza 2009, 521–547.
2 Arthur 2014, 190–202. A sample of the very many philosophers influenced by Leibniz can be found in chapter 3.

3 According to the *Oxford English Dictionary*, the term derives ultimately from the ancient Greek word, οἶκος, meaning 'house' or 'dwelling', with first use in the modern sense of the 'branch of biology that deals with relationships between living organisms and their environment' attributed to the German biologist Ernst Haeckel in his *Generelle Morphologie* (1866). Environmental philosophy, as I understand it, is a wider term that embraces approaches and opinions that are not necessarily also ecophilosophical.

4 See chapter 1, pp. 20ff.

5 Naess 1977, 1989, esp. ch. 7; Sessions 1977, esp. 493ff. The appeal of Spinoza to deep ecologists will be discussed further in chapter 1.

6 Further concerns about Spinoza's philosophy in this context are raised by Lloyd (1980).

7 E.g. Armstrong-Buck 1986 and Gunter 2006. See also Mesle 2008, 76–77.

8 See chapter 3, p. 59.

9 See chapter 3, p. 60.

10 Some of the ways in which Whitehead is indebted to Leibniz are outlined in chapter 3, pp. 59–60.

11 See Sessions 1977, 486, 507.

12 I use 'experiential' and cognate terms here and elsewhere in this book in a very broad sense that should not be taken to imply that experiences are invariably conscious experiences. This use is consistent, I believe, with Leibniz's description of the experiences had by the 'simple monads', i.e. unconscious monads, at *Monadology* §20: 'we experience within ourselves a state in which we remember nothing and have no distinct perception; this is similar to when we faint or when we are overwhelmed by a deep, dreamless sleep. In this state the soul does not differ sensibly from a simple monad' (GP VI 610; AG 215).

13 It has been defended, however, in Phemister 2005.

1 Ecological philosophy
Descartes and Spinoza

> I view philosophy as a basic intellectual inquiry such that the conclusions one reaches within it should properly branch out into a total vision of the world in which man finds himself and of how he should conduct himself there.
>
> (T. L. S. Sprigge, Chair of Logic and Metaphysics, University of Edinburgh, Inaugural Lecture, November 1980: Sprigge 1980, 39)

The 'Study of Nature', writes the English philosopher John Locke (1632–1704) in his *Essay concerning Human Understanding*,

> ... if rightly directed, may be of greater benefit to Mankind, than the Monuments of exemplary Charity, that have at so great Charge been raised, by the Founders of Hospitals and Alms-houses. He that first invented Printing; discovered the Use of the Compass; or made publick the Virtue and right Use of *Kin Kina*, did more for the propagation of Knowledge; for the supplying and increase of useful commodities; and saved more from the Grave, than those who built Colleges, Work-houses, and Hospitals.
>
> (*Essay* 4.12.12)

The sentiment expressed here was by no means peculiar to Locke. The overriding motivation of early modern natural philosophy was the betterment of the human condition, conceived as progress. And undoubtedly the tremendous medical and technological developments made possible by the advances in our scientific understanding of the natural world that the early moderns set in train have, in the developed nations at least, significantly increased human wellbeing in terms of better health and longer life expectancy, more leisure time and other life-enhancing advantages. Nevertheless, such benefits have exacted a high cost. The radical philosophical, cultural and scientific developments of the early modern and Enlightenment periods are often seen as a turning point in humanity's relationship to the natural world. Overturning the old mediæval order and heralding a world-view that continues to inform our understanding of what it means to be human and of the place of human beings in the world, this period – considered by some as the beginning of the Anthropocene[1] – is regarded as the historical root of human alienation from nature that has nourished destructive practices and contributed to current social and environmental crises.[2] As well as the global physical threats to human and nonhuman species from the degradation of the natural environment brought about through the polluting and climate-changing effects of overzealous industrialization and affluent lifestyles, recent research has established strong correlations between our sense of human alienation from the rest of nature and incidences of depression, suicide, juvenile delinquency, petty and violent crime, and other psychological and social disorders that plague the modern world.[3]

Cartesian dualism

Francis Bacon, William Harvey, Thomas Hobbes, Pierre Gassendi, Robert Boyle, John Locke, Robert Hooke and Isaac Newton are just some of the philosophers and scientists whose contributions to the remarkable historical phenomenon known as the early modern scientific revolution utterly changed human understanding of the natural world. Generalizing broadly, the explanation of macroscopic natural phenomena solely in terms of the extension, size, figure and motion – the primary qualities – of elementary particles or corpuscles was the common goal. However, it is Descartes who is most often identified as the key player in the development of the conception of nature as nothing more than a physical machine, devoid of any psychical or mental characteristics. It was he who provided the philosophical justification for the mathematical conception of bodies as measurable and quantifiable, arguing that body is essentially *res extensa*. Extension, or extendedness in length, breadth and depth, is the 'principal attribute' of physical matter (*Principles of Philosophy*, Pt 1, §53: AT VIIIA 25; CSM I 210). Conceived thus, bodies are purely passive objects, devoid of feeling or activity. They function only as mechanical machines, subject entirely to deterministic rules of motion and laws of nature. Animals, trees, plants and even the bodies of humans differ from the inanimate artefacts of human construction only in their origin. The natural machines that God creates no more experience feelings or sensations than do the clocks, pumps and other machines that humans devise. Although Descartes attributes sensations such as hunger and joy to animals, these are given a purely physiological description. Hunger consists in nothing more than a contraction of the muscles in the stomach. Animal joy is interpreted only as the movement of particles from the heart to the brain (Cottingham 1998, 231–233).[4] There is nothing here to suggest that Descartes thought that animals were capable of experiencing the world around them as a world of olfactory, tactile and other sensory qualities.

The removal of conscious or sensory experience from the world of natural things opens the way for fully fledged Cartesian dualism and its separation of rational, human minds from material bodies, including their own. As *res cogitans*, mind is an immaterial, active, free being whose essence or nature is constituted by the principal attribute, thought, or thinking (*Principles of Philosophy*, Pt 1, §53: AT VIIIA 25, CSM I 210). Essentially thinking, perceiving, experiencing and nonextended beings, minds are the exact antitheses of extended, unthinking bodies. This radical incommensurability of mind as thinking substance and body as extended substance entails that the essence of body can be known clearly and distinctly – and truly, since clarity and distinctness were taken as marks of truth – without invoking any reference whatsoever to features of the mind. Conversely, the essence of mind can be known, again clearly and distinctly, without the need to refer to any of the characteristics of body. This in turn forms the basis for Descartes' argument that mind and body are really (not merely conceptually) distinct entities (*Principles of Philosophy*, Pt 1, §60: AT VIIIA 28; CSM I 213). On the assumption that God can create whatever he clearly and distinctly perceives – or in secular terms, on the assumption that whatever is logically conceivable is possible – it follows that God could create minds without also creating bodies and vice versa, could create bodies without creating any minds. In other words, mind and body can, at least in theory if not in actual practice, exist separately from one another.

Not without reason, therefore, did Genevieve Lloyd identify Cartesian dualism as a major contributor to the tragic alienation of humans from nature.[5] Indivisible, immaterial minds are exempted from the mutability and corruption to which divisible, physical bodies are liable, and thus the mind's possible disembodied immortality is secured. Even in the world of nature, where minds coexist in substantial union with their bodies, minds are afforded special status on the ground that they are capable of rational thought and of attaining knowledge of eternal truths and of God. Theoretically, these truths would still be intellectually accessible even if the body normally attached to the mind didn't exist. Indeed, even if the entire material world were to cease, still the mind could in principle discover the eternal truths. The late Val Plumwood considered such 'rationalist rationality' as 'deeply anti-ecological' for it monologically and irrationally refuses to acknowledge the material nature on which it depends. Privileged, but disembedded, Cartesian monological reason is conducted at such a high level of abstraction that it is incapable of engaging with 'the "chaotic" material, bodily, ecological and social order' (Plumwood 2002, 18–20 *passim*).

From the privileging of the rational mind, it is a short, though not inevitable, step to the devaluing of the material and the placing of it beyond the circle of human ethical concern. While some have proposed in recent years that the 'other' be granted moral status precisely because of its 'otherness',[6] Cartesianism veers towards a Platonist devaluation of the physical and a tendency to reduce the natural world to the status of an 'other' whose value is measured only in terms of its instrumental value to the human. Descartes himself evinces a disconcerting ambivalence towards the natural world. His personal preference is towards disengagement with his environment and from the people, animals and other living and nonliving things found there. He confides in a letter to his friend Jean-Louis Guez de Balzac that the 'bustle of the city no more disturbs my daydreams than would the rippling of a stream' (to Balzac, 5 May 1631: AT I 203; CSM III 31) and confesses that when he walks around the streets of Amsterdam, 'I pay no more attention to the people I meet than I would to the trees in your [Balzac's] woods or the animals that browse there' (to Balzac, 5 May 1631: AT I 203; CSM III 31). When he does pay attention, his focus is on the instrumental value that others have with respect to his own comfort and desires. In further confidence to Balzac, he confesses to the pleasure he feels from seeing his own needs being satisfied.

> Whenever I reflect upon the doings of passers-by I get the same sort of pleasure as you [Balzac] get when you watch the peasants tilling your fields, for I can see that all their work serves to enhance the beauty of the place I live in, and to supply all my needs. Whenever you have the pleasure of seeing the fruit growing in your orchards and of feasting your eyes on its abundance, bear in mind that it gives me just as much pleasure to watch the ships arriving, laden with all the produce of the Indies and all the rarities of Europe.
>
> (To Balzac, 5 May 1631: AT I 203–204; CSM III 31–32)

This letter paints a picture of Descartes as a singularly solitary figure, preferring his own company to that of his fellow human beings or animals. There is nothing amiss in that, but the letter also hints at a less than salubrious self-centredness in Descartes' delight in the activities of other human beings on account of the commodities their efforts provide for him. The same attitude is extended even to nature's country streams and secluded valleys. These too, it seems, are valued for their utilitarian capacity to safeguard Descartes' solitude by making even 'the greatest talkers fall into reveries' or 'transport them into ecstasies', for whether in reverie or ecstasy, the would-be vociferous fall silent (to Balzac, 5 May 1631: AT I 203; CSM III 31).

Nevertheless, Descartes was fascinated by the mechanical operation of natural processes, seeking out the secrets hidden in matter and publishing the results of his enquiries in scientific treatises, such as *The World, The Optics, Treatise on Man, Description of the Human Body* and *The Passions of the Soul.* The same delight in exploring the intricate workings of the natural world is found in Descartes' follower, the Oratorian priest Nicolas Malebranche (1638–1715), who found in nature evidence of the wisdom and omnipotence of God. '[T]hose who have used only their eyes', he wrote, 'have never seen anything so beautiful, so fitting, or even so magnificent in the houses of the greatest princes as what can be seen with magnifying glasses on the head of a simple fly' (*Search after Truth,* Bk 1, ch. 6: Malebranche 1979–92, I 64; LO 31). All the same, neither Descartes nor Malebranche offers us any explanation as to why God created the physical world at all. Descartes, for instance, merely asserts that 'Nature' teaches him that he has a body with which his mind is substantially united and that God would not have deceived him with respect to his belief that his sense perceptions are representations of an actually existing world (Meditation 6: AT VII 80–81; CSM II 55–56). However, neither of these assertions explains why God did not create only minds and their perceptions, without also creating any actual material bodies to which their perceptions correspond. God could have ensured that 'Nature' taught Descartes that bodies were merely figments of his imagination. All in all, the existence of bodies appears rather futile in the philosophies of Descartes and Malebranche. Both contend that minds alone are capable of experiencing pain or pleasure. Nonhuman animals, lacking souls, are only physical machines: arrangements of extended, figured, moving parts, more complex than, but analogous to, the machines that humans design and construct. Since animal machines feel no more pain that does a clock, we may with clear consciences, investigate and dissect them even as their hearts beat and their bodies move.[7]

The Cartesian picture of the natural world was taken forward in the latter half of the seventeenth century by Locke. In his hands, the distinction between the primary qualities of extension, figure, shape, motion and bulk of elementary particles and the phenomenal secondary qualities of bodies, such as their colour, feel, odour, sound and taste, was philosophically grounded, with explanation of the latter in terms of the former established as the most epistemologically viable hypothesis.[8] Although he himself espoused dualism, Locke's philosophy was instrumental in the formation of eighteenth-century materialism both in Britain and France.[9] The basic Cartesian–Lockean model by which natural phenomena are to be explained by reference to the motion, number, size and shape of these interacting extended, solid, but essentially lifeless, particles remains with us today, whether it be in physicalism's reductionist attempts to understand consciousness entirely in terms of the chemistry of the human brain or the physical and biological sciences' reductionist attempts to understand life as an emergence from nonliving, nonexperiencing molecules and subatomic or elementary particles. And just as Descartes showed little inclination to try to imagine how the world might appear from the perspective of the 'passers-by' in the streets of Amsterdam, so too, barring some notable recent exceptions,[10] physicists and other natural scientists even today appear disinclined to contemplate the possibility that there might be something that, to coin a phrase from Timothy Sprigge and Thomas Nagel,[11] 'it is like to *be* an atom, an electron, a quark …', to experience the world from those perspectives.[12]

One might have expected that materialist philosophies, placing everything on the body, would value the physical above all. However, with the expulsion of the psychical from ground-level ontology, the human body has come to be viewed as a purely physical machine and the task of explaining our minds' emotions, value judgments, spiritual and aesthetic sensibilities taken up by neuroscientists working on the assumption that the psychical dimensions of our being are in principle reducible to physical states of the body or of the brain. The mind's conscious thoughts, emotions and sensations, its religious and aesthetic sensibilities, and so forth are seen as emerging, in a manner not yet fully understood, from the physical complexity that is the microphysical world of insensate matter. The richness of experienced life in all its social and spiritual complexity is thus subordinated to the physical and material that is consequently disenchanted and under threat of devaluation on account of its lifelessness.[13]

As an antidote to the Cartesian dualist and materialist–physicalist perspectives, ecological philosophers have been encouraged by earlier panpsychist philosophies, such as those expounded by Spinoza and Whitehead, that attribute equal ontological status to the physical as to the psychical. For the remainder of this chapter, we pay tribute to Spinoza and consider the influential role of his philosophy in the work of deep ecologist, Arne Naess.[14]

Spinoza

Spinoza's *Ethica ordine geometrico demonstrata* (simply referred to as the *Ethics*)[15] presents a system of speculative philosophy demonstrated 'geometrically' from real definitions, self-evident axioms and postulates.[16] The *Ethics* defends the ontological position now known as absolute monism, the view that Reality is One. In Spinoza's hands, this is expressed as the view that there exists only one true unchangeable 'Substance'.[17] He establishes the identity of this Substance as God early in the *Ethics* (*Ethics* I, P14). Only later in the work do we find his controversial identification of God with Nature (*Ethics* IV, Preface).[18] Spinoza's capitalization of these key terms is important. It signals that Reality is a single indivisible infinite and perfect whole and that the things that we ordinarily and incorrectly regard as substances – individual living things and inanimate objects – are not isolated and potentially separable parts of the whole. '[T]he eternal and infinite essence' of God/Nature is constituted and expressed by the divine attributes, the entire essence wholly expressed by each.[19] It is unclear from Spinoza's texts just how many attributes God possesses.[20] Perhaps there are infinitely many, but Spinoza identifies only two: thought and extension. The particular entities we regard as making up the natural world are collections of 'modes', 'affections' or 'modifications' (the terms are in this context equivalent) of these attributes. In Spinoza's technical language, 'mode' signifies 'the affections of a substance, *or* that which is in another through which it is also conceived' (*Ethics* I, D5: Gebhardt II 45; Curley 409). Expressed less technically, we can understand the term as referring to the differing states or 'ways of being' that a substance can assume. So, when Spinoza's Substance, God or Nature, is considered as an extended substance, its modes are the myriad individual bodies of different shapes and sizes, sometimes moving, at other times at rest, that comprise the whole. From the other side, considering this same Substance as a thinking substance, its component modes are the myriad ideas or individual experiences, such as those we know as our own lived experiences and those experiences that comprise the lives of other living things.

Thus, as a thinking substance, that is, as Substance whose essence is expressed and constituted by the attribute of thought, Spinoza's God is modified, or gives rise to, infinitely many ideas, some of which – namely those we have of our own bodies – collectively constitute our minds. Any particular human mind is the set of all the ideas that constitute that mind's mental states or experiences over its entire lifetime. These ideas or experiences are effectively ideas of its own body (*Ethics* II, P13), that is to say, in the first instance at least, they are ideas of the body through which the mind encounters the world as a physical reality. Meanwhile, as an extended substance, that is, as Substance whose essence is expressed and constituted by the attribute of extension, Spinoza's God is modified, or gives rise to, infinitely many individual bodies or modes of extension, some of which make up those very bodies we experience as our own, with all the diverse configurations, sizes, shapes and motions that the geometrical attribute encompasses and indeed makes possible.

As the only Substance in possession of attributes, Spinoza's God is the true and only source of activity in the universe. The God of Spinoza's metaphysics is both the only thinking substance and the only extended substance. This God is *natura naturans*, literally, 'nature naturing'.

> By *Natura naturans* we understand a being that we conceive clearly and distinctly through itself, without needing anything other than itself ... i.e., God.
>
> (*Short Treatise on God, Man, and His Well-Being*:
> Gebhardt I 47; Curley 91)[21]

In contrast, the particular ideas that God thinks and the particular body-modes that God 'extends' comprise the natural world. This is *naturata naturata* or 'nature natured', that is, all the particular ideas and bodies that follow from God's naturing activity:

> by *natura naturata* I understand whatever follows from the necessity of God's nature, *or* from any of God's attributes, i.e., all the modes of God's attributes insofar as they are considered as things which are in God, and can neither be nor be conceived without God.
>
> (*Ethics* I, P29 Scholium: Gebhardt II 71; Curley 434)[22]

Because God's attributes of thought and extension each express, in their own ways, one and the same essence, namely, the essence of God, an exact parallelism holds between the modes of thought and the modes of extension that comprise the particular things in *natura naturata*. For every idea-mode that arises from God insofar as God is thinking Substance, there is a corresponding body-mode that arises from God insofar as God is extended Substance. The logical connection of ideas matches exactly the sequence of causes and effects among physical things: 'The order and connection of ideas is the same as the order and connection of things' (*Ethics* II, P7: Gebhardt II 89; Curley 451). As Spinoza goes on to explain, the connections between ideas are comprehended only through the attribute of thought, while the connections between things are explicable only through the attribute of extension. But in each case, that which is being explained is the same. The attributes only provide, as it were, the respective explanatory framework.

> [A] circle existing in nature and the idea of the existing circle, which is also in God, are one and the same thing, which is explained through different attributes. Therefore, whether we conceive nature under the attribute of

Extension, or under the attribute of Thought, or under any other attribute, we shall find one and the same order, *or* one and the same connection of causes, i.e., that the same things follow one another.

(*Ethics* II, P7 Scholium: Gebhardt II 90; Curley 451)

The result is a thoroughgoing panpsychism. For every body conceived under the attribute of extension there is a corresponding idea under the attribute of thought. Each particular thing is conceived under both attributes, not as a dualistic conjunction, but as one and the same thing understood in two different ways. Conceived under the attribute of extension, a human being comprises body-modes that combine in a nested structure to form the complete human body and which, extended also over time, includes all the successive physical states (motions and resistances and interactions with others) of the whole body throughout the entire period of its existence, from conception to death. Under the attribute of thought, on the other hand, a human being is a mind comprising the entire collection of ideas that represent that particular human body,[23] again including ideas of all the nested parts and of each of the body's physical states over time. The relation of idea-modes to body-modes is not unique to humans, but holds for every individual thing (*Ethics* II, P13 Scholium). Other sets of idea-modes arising from God's attribute of thought constitute the conscious- or feeling-centres of dogs, cats, mice and all other animals or biological entities, right down to simple amoeba and beyond.[24]

For the things we have shown so far are completely general and do not pertain more to man than to other Individuals, all of which, though in different degrees, are nevertheless animate. For of each thing there is necessarily an idea in God, of which God is the cause in the same way as he is of the idea of the human Body. And so, whatever we have said of the idea of the human Body must also be said of the idea of any thing.

(*Ethics*, II, P13 Scholium: Gebhardt II 96; Curley 457)

Spinoza postulated a world of simplest bodies that are combined into larger composite bodies. These in turn enter into more structurally complex bodies, composites of composite bodies. Increasingly more complex bodies are composites of these composites of composite bodies, ad infinitum. Thus, for instance, the cells of our own bodies combine to form the internal organs, sense organs, the blood, the brain and so forth, and all these together forming the whole human body. Our own bodies, he imagined, are themselves parts of larger organic structures, and these in turn of even bigger organic bodies. Ultimately, all reside in the organic body that is the whole physical world: the infinite body-mode that fully expresses God's essence under the attribute of extension and that contains within itself all finite body-modes, all finite bodies.[25] Meanwhile and in parallel, under God's attribute of thought, the same structure is manifested in the logical connections between ideas (*Ethics* II, P7), culminating in the idea that God has of himself. Contained within this idea are the constituent ideas – the ideas of the simplest bodies, the ideas of the composite bodies, of the composites of composite bodies, and so forth. Thus are all the ideas of each individual body, from the smallest microscopic particles to the larger mammals and beyond, contained and connected within the overall infinite idea of the world as a complete, infinite whole.

Each individual thing has its own *conatus* or power by which it strives to preserve its own being in existence. From the side of ideas, the more adequately or truly an idea represents the whole, the more it incorporates the logical connections to other ideas within the whole, and the greater its being and power. For instance, to use Spinoza's example, an adequate idea of the worm in the blood would relate the worm not only to the particles in the blood that surrounds it, but would also understand the blood particles as parts of a larger organism, and the relations of this organism to others, and so forth.[26] Ultimately, the fully adequate idea of the worm would identify the worm in all its particularity and situatedness within the whole and would demonstrate why, given the necessity of the connections between all things that follow from the essence of God, the worm is, and must be, just as it is. Of the ideas that make up the human mind, most are inadequate. Insofar as they consist of fleeting sense impressions, hearsay reports and unfounded suppositions, they do not carry with them the evidence of their truth.[27] They may empirically inform us *that* certain things happen, even that they are in the habit of happening regularly, but they do not show us *why* they occur. They give us no knowledge, for instance, of an event's necessary causes and therefore offer no explanation of why it happened in the way that it did, nor any reason to believe that it will happen again, even in similar circumstances. Thus, inadequate ideas – ideas that belong to what Spinoza calls the 'knowledge of the first kind, opinion or imagination' (*Ethics* II, P40 Scholium 2: Gebhardt II 122; Curley 477) – are not particularly effective as tools to assist the individual to preserve its own being.

Spinoza identifies a further two kinds of knowledge: knowledge of the second kind and knowledge of the third kind. The adequate ideas involved in these kinds of knowledge actively reason about the necessary causes of things. Knowledge of the second kind is general knowledge of the 'common notions' of things (*Ethics* II, P40 Scholium 2: Gebhardt II 122; Curley 478). It allows us to classify particulars into classes of things based on their essential properties. It is, in essence, objective scientific knowledge. But as such, it still places the knower outside the thing that is known. Only when adequate ideas fall under the third kind of knowledge is knowledge of things in all their particularity and in their unique relations to other modes within the whole attained. These relations include, of course, the relation of the thing known to the knower, as well as the knower and the object's relation to the true cause, God. Self-knowledge, then, is clearly of the utmost importance. The more we come to know ourselves, the more we come to understand ourselves in relation to the whole, to appreciate the role we play within the whole, and to comprehend the reason or cause of all ideas – God. When by the third kind of knowledge, adequate ideas recognize God as the ultimate source of everything and realize that things are the way they are by the necessity of God's essence, these ideas and their objects are conceived *sub specie aeternitatis*, as eternal modes of God's attributes. When the human mind, as the idea of the human body, conceives its body 'under a species of eternity', the eternality of the idea is, in effect, the eternality of the mind. Thus, 'our mind, insofar as it involves the essence of the body under a species of eternity, is eternal' (*Ethics* V, P23 Scholium: Gebhardt II 296; Curley 608).

A human mind just is those ideas in God that are ideas of its own human body. The more adequate are these ideas, the more this mind is able to preserve its own existence, for the more adequate its idea of its body, the more this idea incorporates the ideas of the physical causes of its body and the more the physical causes themselves are efficacious. The mind, in effect, becomes the clear logical sequence of ideas that demonstrate why things are as they are and why they could not be otherwise. Since it is the idea itself, the mind itself becomes the logical reason for the continuation of its body. It contains the causes within itself and the more it does so, the more active and causally efficacious it becomes and the less it is acted upon from outside.

The activity of the ideas is represented under God's attribute of extension as the activity and the complexity of the human body, the object of the mind's ideas. The more adequate the idea of the body, the more active both the idea and the material correlate of that idea. God, of course is the limiting case – the *causa sui*, the ultimate reason for everything. God's omniscience and omnipotence are inseparable, even identical. God is pure activity – God's essence is fully active.

> God's power is nothing except God's active essence. And so it is as impossible for us to conceive that God does not act as it is to conceive that he does not exist.
>
> (*Ethics* II, P3 Scholium: Gebhardt II 87; Curley 449)

In this lies the key to understanding why, within Spinoza's philosophical system, the best interests of all rational minds, the surest means of attaining self-preservation, is to strive towards the fully adequate idea of God that is God's eternal idea of himself, the perfect idea of the whole. The more an idea (or mind) includes the idea of God's essence, the more active and causally efficacious the idea. The more the idea contains its causes or reasons (God and the modes of God's attributes), the more the idea is self-caused rather than passively acted upon and the more freedom the individual idea/mind exercises.

To strive to know God is also the path to virtue, salvation and blessedness. The idea of God is the idea of the perfect being. Knowledge of perfection arouses feelings of pleasure and joy in the perceiver and is accompanied by love of the cause of such pleasure in the perceiver. In this way, the third kind of knowledge leads necessarily to an intellectual love of God, to a love of God not as an imagined present being, but as truly eternal (*Ethics* V, P32 Corollary). The mind's intellectual love of God is the love of that which God loves. Spinoza equates God's love of himself with God's love of men: 'insofar as God loves himself, he loves men, and consequently ... God's love of men and the Mind's intellectual Love of God are one and the same' (*Ethics* V, P36 Corollary: Gebhardt II 302; Curley 612). Of course, God's idea of himself also includes ideas of all the other idea-modes and body-modes that follow from God's attributes. It contains ideas of all that comprises *natura naturata*. These too are surely included within the range of God's love of himself. In the ecological context to which we turn now, God's love of himself is not restricted to the love of 'men', but is equated with the love of *all* living things, love that is directed towards everything in *natura naturata*.

Ecosophies and the deep ecology movement

Arne Naess is not alone in recognizing the ecological potential of Spinoza's philosophy,[28] but it is in the writings of Naess that we find this potential exploited most fully. Drawing on Spinoza and adopting a basically Kantian position Naess endeavoured, not to find a set of universal presuppositions that ground the very possibility of experience itself, but instead to offer new presuppositions that would change those experiences. Believing that our actions and attitudes are intimately related to, and indeed arise from, our deeply held, but often un- or underacknowledged metaphysical beliefs, Naess instinctively grasped that if we are to advance beyond a shallow ecological and piecemeal reactionary response to particular environmental degradations and the progressive depletions of natural resources, we must fundamentally re-envision ourselves in relationship with the natural world, changing our metaphysical conceptions both of ourselves and of the world we inhabit. We must conduct a thorough examination and where necessary a radical overhaul and replacement of our most fundamental presuppositions, for it is only, Naess claims, when our beliefs are grounded in considered and deeply held metaphysical positions that effective and sustained behavioural change can become a reality. Rather than expose, as Kant had done, the universal categories and intuitions that make phenomenal experience possible, Naess reinstated a pre-Kantian metaphysical project with a Kantian twist. Rejecting the notion that philosophy should aim to arrive at a universally agreed true metaphysical account of reality and abandoning Kant's project to find the presuppositions that underlie *all* human experience, Naess nevertheless acknowledged that our experiences of the world, our attitudes towards it, and the decisions and actions that arise from these are grounded in our beliefs about the nature of reality. At the same time, recognizing and indeed celebrating human differences, Naess realized that we are not all predisposed towards or inspired by the *same* metaphysical presuppositions. He therefore encouraged each one of us to develop our own ecological metaphysical systems – our own individual ecosophies – that will enable each of us to make the lifestyle and attitudinal changes required for us to live at peace in nature and meaningfully address and redress global and local environmental concerns.

It is important to distinguish carefully the highly personalized ecosophies of individuals from the basic principles or platform of the deep ecology movement (DEM) to which all DEM supporters are committed. In a 1972 lecture,[29] Naess characterized (and simultaneously founded) DEM in terms of seven 'norms or tendencies': the adoption of a 'relational total-field image'; the acceptance, in principle, of the principle of biospherical egalitarianism, and of principles of diversity and symbiosis; the adoption of an 'anti-class posture'; the commitment to 'fight against pollution and resource depletion'; recognition of complexity and of its value; and support for 'local autonomy and decentralization'. In 1984, Naess, together with George Sessions, drafted a proposal of a closely related set of eight 'basic principles' comprising the 'platform' of DEM.[30] The first two principles made more explicit the relationships between the principle of bio- spherical egalitarianism and the noninstrumental, intrinsic value and flourishing of human and nonhuman life and re-emphasized the importance of diversity and complexity. Intended as a set of principles upon which members of DEM agree, the 1984 proposal also reminded members of their obligation to act, whether directly or indirectly, to save the environment both locally and globally. The proposal also stressed the need to insist upon technological, economic and ideological policy changes to promote ecologically sustainable economic growth and the introduction of 'soft' technologies compatible with cultural diversity. It also advocated a refocusing of attention away from the notion of 'standard of living' to 'quality of life' and highlighted our human obligation to restrict the limitation of ecological diversity to cases where it is required for the satisfaction of 'vital' human needs.[31] Thus, the proposal promoted the preservation of wild- erness areas free from human interference and suggested that human population growth must be controlled in order to allow the nonhuman to flourish.

Naess firmly believed that the DEM principles or platform must be meta- physically grounded by distinctive personal ecosophies thoughtfully developed by individual supporters of DEM. It is not required that all DEM supporters share the same ecosophy. Although Naess and Sessions maintained that the 'basic principles within the deep ecology movement are grounded in religion or philosophy', they also understood that there are many possible routes to the same end: there is no single 'definite philosophy or religion among the supporters of the deep ecological movement', for 'there is a rich manifold of fundamental views compatible with the deep ecology principles' (Naess and Sessions 1984, 3). And so, while every personal ecosophy should identify 'ultimate premises' capable of justifying the principles of DEM (Naess 1988), these premises and their religious and philosophical contexts need not be the same across all ecosophies. As Naess pointed out, the same conclusions may be proven in dif- ferent ways: 'very similar or even identical conclusions may be drawn from divergent premises. The principles (or platform) are the same, the fundamental premises differ' (Naess and Sessions, 1984, 3).

Ecosophy T

Naess's own ecological philosophy – Ecosophy T[32] – is an ecological reworking of Spinoza's system of philosophy. While not unqualifiedly successful,[33] there is nevertheless no doubt that Spinoza's philosophy does provide ultimate premises that strongly support the norms and principles of DEM. Naess himself highlights Spinoza's conception of the individual or particular as contained 'in' or as an integral 'part' of the whole as especially significant (Naess 1969, 82). Spinoza's conception of *natura naturans* as an organism comprising a complex arrangement of smaller organic part-wholes with decreasing degrees of complexity describes a universe that is both infinitely complex and infinitely diverse: infinitely many modes follow from God's essence in infinitely many ways. In keeping with the DEM norms and principles, this complexity is uncomplicated. Spinoza's complex and diverse world has an orderliness grounded in the simplicity of God. Diversity without order would simply be chaotic – and complicated. Diversity with order is not complicated, but it is complex and, crucially, it is life-supporting. Spinoza's *natura naturata* is a world of symbiotically interdependent particular things, all intrinsically related, logically and causally, to each other and to the whole. Symbiosis does more than simply acknowledge that, in an interconnected universe, we need other creatures in order to survive and thrive; for Naess, as for Spinoza, it is also a clear indication of our absolute inseparability from others. Disruption of our symbiotic relations to others is a destruction of our very identities. No organism can remain what it is when its relations to others are changed. In this, it serves also as important metaphysical underpinning of a relational conception of individual identity (Naess 1969, 83), a key feature of Naess's 'relational total-field image' norm.

The concept of a total-field image encapsulates this all-pervasive universal interconnectedness.[34] Following Spinoza, Naess accepts relational qualities that are intrinsic relations, that is to say, relations between things that make a difference to the very natures or identities of the things related. When things stand in intrinsic relations to each other, the relata are changed by the fact of the relation. When *A* and *B* are intrinsically related to each other, both are changed by the relation (Naess 1989, 28). Their very identities are defined and constructed by the relations.[35] Intrinsic relationality highlights the fact that our very being depends upon the being of others. Were a mouse, Naess illustrates, placed in a vacuum, 'it would no longer be a mouse' (Naess 1989, 56). Like the mouse and other organisms, each human person exists as a 'relational junction within the total field' (Naess 1989, 56). Organisms are not individual substances existing in environments from which they could in principle be separated without loss of their identities. For Spinoza, organisms are idea-body modes identified by their position vis-à-vis the other modes within the integrated whole; for Naess, they are 'knots in the field of intrinsic relations' (Naess 1989, 28).

These knots endeavour to preserve themselves. Each part of the whole possesses a *conatus* or drive towards self-preservation. In humans, as we saw in the brief outline of Spinoza's thought offered above, self-preservation is best served by striving to increase the adequate ideas of the second and third kinds of knowledge, especially the third kind, the intellectual love of God. This aspect of Spinoza's holistic panpsychist metaphysic finds expression in the most basic norm of Naess's Ecosophy T: 'Self-realization!' or wide-identification (Naess 1989, 197). Our minds are idea-modes. By increasing the adequacy of the ideas that constitute our minds, becoming aware of the connections between things as we increase the second and third kinds of knowledge, we come to appreciate how our own identities are intricately entwined with our relations to everything in the world. We come to understand how the whole universe (*natura naturata*) is essential to the identity of our individual selves. In other words, we extend the notion of our individual selves to include others within its scope. Indeed, the more we understand ourselves as dependent upon the other modes under God/Nature's attribute of thought, the more these others are consciously acknowledged, not as distinct from ourselves, but as parts of ourselves. Our selfhood is expanded to include more and more of the wider whole, making more explicit the logical and causal connections between the parts. In the nondualistic third kind of knowledge in which knower and known are fully integrated, the idea or group of ideas that is the human mind expresses the human body as existing within, and as one with, the larger whole. In this state, the mind fully understands that everything in the world is interrelated and interdependent and that nothing can exist separated from this network of modes. Striving towards the identification of ourselves with the whole of the world or universe – striving to become in Spinozistic terms, the infinite mode by which God knows himself – we lose our sense of alienation from external things. In return, we gain a sense of belonging and empathic connection to the modes in God that constitute living things other than ourselves.

This deeper sense of the interconnections of all with all leads to an appreciation that the good of each particular is actually the good of all. Each part within this universal whole is indispensable to the wellbeing of all the other parts. Every organism is an integral, indispensable part of the whole, with its own unique and intrinsically necessary role within the organization of the interconnected whole. The world as a whole functions well only if its parts are healthy. Similarly, the organic parts are healthy only if their parts in turn function efficiently. As well as a brain, a mammal needs to have eyes, ears, mouth, lungs, heart, nervous system and so forth; a plant must have flowers, stamen, leaves, sap, etc. In all cases, it is imperative for the organism as a whole that each of its component organisms operates to the very best of its capabilities, i.e. that each performs its own unique function well.[36]

Spinoza had observed that the perception of the happiness of others produces active emotions of joy and pleasure in the perceiver and, furthermore, that these joyful and pleasurable emotions are accompanied by feelings of love in the perceiver, love that is directed towards the perceived cause of the pleasurable experiences. The agreeableness and love we feel towards people in whose company we feel comfortable is a common and natural phenomenon and, except in extreme cases of irrational psychotic illness, the love we feel towards others is a desire for the wellbeing and happiness of the person loved for their own sake, and not on account of the pleasure it brings to the lover. Such love is the foundation of true friendship, which consists in the pursuit of the happiness of the friend, not because it increases one's own joy (although it does), but for its own sake. As we noted earlier, love finds its highest object in the perception of the perfection of God and the love of that which God loves. Extended from human minds to the rest of *natura naturans*, the intellectual love of God is expressed as the love of all that follows from God's essence, the love of all idea–body modes of God's attributes.

In this way, Spinoza's philosophy can be adjusted to serve the DEM platform's biospherical egalitarianism norm. Everything in Spinoza's universe is a valued member of the whole, while the whole itself is perfect, infinite and intrinsically valuable.[37] The parallelism Spinoza envisaged between ideas and bodies implies that every body is a living body, in the sense that for every body, there will also be ideas (ideas of its body) that serve as the 'feeling-centre' of its being. Life is all-pervasive in the Spinozistic world. Accordingly, in Naess's hands, bioegalitarianism is spared the need to justify prioritization of the living over the inanimate, for it is really a '[b]iospherical egalitarianism – in principle', an ecoegalitarianism that extends beyond the technically biological to encompass such things as ecosystems and cultural communities (Naess 1989, 28–29). Spinoza's parallelism of ideas and bodies also avoids any hint of Cartesian prioritization of mind over body. For Spinoza, minds (as collections of idea-modes) and bodies (as sets of smaller body-modes) are, as it were, the 'same-but-different': the same insofar as they are all modes of God, but different insofar as the first are considered as modes of God's attribute of thought and the second as modes of God's attribute of extension. Nor is there any reason to prioritize one mode over another of the same type: one idea-mode is just as valuable as any other idea-mode for the perfection of the whole, and any one body-mode just as valuable as any other body-mode. As far as possible, then, psychophysical organisms, as perceiving-embodied things, are to be treated with equal respect as ends in themselves, all striving to preserve their own being.[38]

That every individual's *conatus* strives to preserve its own being is just one of the ecologically important aspects Naess finds in Spinoza's thought (Naess 1969, 87). Another is Spinoza's account of how the adequate ideas that characterize the second and third kinds of knowledge increase the perfection of rational beings, especially at the higher levels where they give rise to feelings of joy, freedom and the ability to 'partake in eternity' (Naess 1969, 86–100 *passim*). All adequate ideas are active ideas, but they are most active when they are adequate ideas of the third kind of knowledge – that is, when they are ideas of particulars in relation to the perceiver and to the whole, when they are intellectual intuitions that approach the kind of idea that God has of himself. These ideas do not separate the knower from the known. Nor do they separate the understanding from the will. Moreover, Spinoza's parallelism of mind (or ideas) and body means that the activity of the mind is always found together with the activity (motion) of the body. The intellectual love of God that characterizes the third kind of knowledge, therefore, automatically leads to virtuous action that promotes the wellbeing of all, for the perception of the perfection of God and of all that follows (necessarily) from God stimulates the rational being to act lovingly in accord with God's understanding/will, to care for *natura naturata*, and as far as possible to be the cause of joy and happiness in others.

In keeping with Spinoza, Naess maintains that deeply reasoned (or intuited) metaphysical views automatically lead to action. Practical actions, under a Spinozistic analysis, arise naturally from deeper metaphysical convictions. Accordingly, at the core of Ecosophy T is the notion of activity that arises naturally from considered reflection. Naess's conception of activity is more than mere action. It is 'praxis': action (or practice) informed by theory. Theory, properly understood, leads inevitably to action. On this point, Spinoza's view enables Naess's Ecosophy T to serve as metaphysical grounding to DEM platform norms that encourage members' active engagement in environmental movements seeking political and cultural change. Spinoza's notion of activity lies behind Naess's and Sessions' injunction to sociopolitical action, as for instance in the DEM platform's fourth principle (anti-class posture). In conjunction with the principle of biospherical egalitarianism and Naess's expansion of Kant's maxim into one that prohibits treating any living being only as a means to an end (Naess 1989, 174), the DEM anti–class posture principle promotes action against all forms of exploitation. Other platform principles, such as the seventh (local autonomy and decentralization) positively encourage the treatment of others as Kantian ends in themselves, but above all, DEM stresses that all political action aimed at the introduction of environmental strategies to promote, for instance self-sufficiency, reduced food miles, or the simplification of government structures, be conducted peacefully.[39]

Although Naess attached great ecological value to the active dynamic character of Spinoza's concept of Nature (*natura naturans*) (Naess 1969, 81, 91ff.), it is questionable whether Spinoza's active God or Nature really is well-suited to underpin the kind of action required of DEM activists. As *natura naturans* (nature naturing), Spinoza's God is the only truly active Substance and the sole source of all activity. As *natura naturata* (nature natured), the world itself is inherently passive. The activity in *natura naturata* evidenced in the motion of bodies[40] and in the mind's conscious and self-conscious adequate ideas is primarily God's activity, for particular things in the world exist only as idea- and body-modes unfolding from God's attributes of thought and extension. As modes not substances, individual organisms do not act as individual agents; they lack the inner active force needed to ground true agency. For Spinoza, the true agentic force resides in God and consequently what we like to regard as our own agency is not so much ours as God's. As George di Giovanni remarked recently:

> Spinoza's system according to Jacobi, undermined the possibility of human agency because it compromised the individuation of any presumed subject of action. Since it reduced the identity of any such subject to a mere semblance, it made impossible the attribution of action – at least, in a way that would make the subject truly responsible for it.
>
> (di Giovanni 2011, 17)

Even if there is a way of attributing a sufficiently strong sense of individual agency to Spinoza's modes, neither their behaviour nor God's could be different from what it is. Were God to be different, then God would not be this God, but a different God. But no other God or Substance is possible (*Ethics* I, P14). This God exists necessarily. What follows from this God's essence is also strictly necessary. The individual modes that make up *natura naturata* follow from God's essence in a definite and determinate fashion. In a system in which no other gods or worlds are possible, past, present and future follow with an absolute inevitability. Nothing can happen in any other way than it does (*Ethics* I, P33).

Despite this brutally strict necessitarianism, Spinoza did offer a rationalist account of freedom and power in terms of adequate ideas of the third kind of knowledge that is compatible with his unyielding denial of contingency. The free person who intellectually loves God knows that God is the true cause of everything and strives to understand the eternal natures of things, to understand them *sub specie aeternitatis*. In this lies the reason why the free person does not fear death (*Ethics* IV, P67). The free person is aware that there is a part of the mind that is eternal (*Ethics* V, P23). This is the part that knows and loves God intellectually through the third kind of knowledge (*Ethics* V, P33). The free person desires nothing more than to know and love God through loving others and that they too may love God intellectually and experience the virtuous joy and salvation this entails. Nevertheless, such freedom is still entirely determined by God's essence. We are not free to determine whether we will be free. If we are free, then we have power to bring about what we desire or will, but we have no power to make ourselves free if it is not already determined by God's essence that we will be so. Indeed, the knowledge pursued in the third kind of knowledge exemplified in the free person just is the knowledge of the absolute necessity of things, an understanding of why things are as they are and why they could not be otherwise.

Spinoza's necessitarianism has potentially serious implications for the practical aspects of the DEM platform. Are we determined absolutely to act in the way we do? The answer for Spinoza must be affirmative. Does this mean that it is impossible for human actions to divert the future course of events from its necessitated path? Certainly, Spinoza's metaphysics implies that climate change, species extinctions, deforestation, rising sea levels, and other environmental challenges arise, like everything else, with logical and causal necessity directly from the divine essence. However, this does not necessarily mean that we are impotent to change the direction in which current events seem to be headed, for though the environmental crises we face are necessitated by the divine essence, so too are the actions we take to ameliorate and resolve them. Naess's calls for sociopolitical action were just as necessitated by the divine essence as everything else. And although from our limited perspectives we do not know what the future will be, we can be assured that whatever comes to be in the future, our actions in the present will have played an essential part in the logical and causal sequence of events that led up to it.

Unfortunately, there is another more serious difficulty associated with Spinoza's conception of God or Nature and the necessity that this entails: the complete absence of teleology in Spinoza's system. Spinoza's God does not act by final causation in pursuit of goals or ends that God understands as good. On Spinoza's view, were God to act in pursuit of an end or goal, there would be something that God lacked. But God, being perfect, lacks nothing (*Ethics* I, Appendix: Gebhardt II 80; Curley 442). God cannot therefore act to bring about any state of affairs that is not in some sense already present in God. It follows therefore that the optimism that Naess discerns in Spinoza's thought is misplaced. Spinoza's world is not in any sense progressing towards a future state better than the present; all that changes is that we come to realize by the third kind of knowledge that time itself is an illusion and that past, present and future have always existed *sub specie aeternitatis* in a single eternal 'present'.

The perfection of God also presents challenges for a Spinozist ecosophy, for it suggests not only that God is perfect, but also that everything that follows necessarily from God is perfect too. No matter how imperfect it may seem from our perspective, the world that arises necessarily from God's Nature is as perfect as God himself. '[A]ll things proceed by a certain eternal necessity of nature, and with the greatest perfection' (*Ethics* I, Appendix: Gebhardt II 80; Curley 442). The world, it would seem, is already perfect and stands in no need of improvement. From this perspective, what appears to us as ecological damage cannot, from within Spinoza's system, be considered as right or wrong, good or bad, or in need of restoration. Claims about goodness or badness can only be made from a limited point of view. From the absolute perspective of God, talk of good and bad, of the desirable and undesirable, is inappropriate (*Ethics* I, Appendix). Nor can our own desires and actions be seen as anything more than the logical and causal outcome of preceding events. Accordingly, Spinoza's ethics offers only a descriptive ethical naturalism that outlines the characteristics of the virtuous person. Prescriptive advice about how we should or could act is out of place in a world where nothing can be other than it is. Similar points arise in relation to ascriptions of beauty. Strictly speaking, just as nothing is good or bad in the world, neither is anything beautiful or ugly. As James Morrison has indicated, perceptions of beauty are imaginary and subjective: '[i]n itself or in relation to God, nothing is beautiful or ugly' (Morrison 1989, 360).[41] If we were to claim, *pace* Spinoza, that *natura naturans* and *natura naturata* are beautiful, it could be only in the way that mathematics and logic, not sensed objects, are beautiful.[42]

Each of these concerns about individual autonomy and freedom, contingency, teleology, perfection, ethics, aesthetics and optimism arises directly from Spinoza's absolute monism. His monism even poses a threat to individuality per se and undermines the ecological attractiveness of Naess's notion of Self-realization or wide-identification. The wholeness and substantial singularity of the world threatens to occlude the identity and autonomy of individual things. It might be argued that the perfecting of the Spinozan rational being lies in transcending his or her particularity, not in strengthening it, or, to put it in another way, relinquishing the individual self in favour of the universal Self. The perfecting of Spinoza's free man consists in striving as far as is possible to understand the world *sub specie aeternitatis*, from the divine objective viewpoint, not from one's own situatedness within a particular environment. If immortality is to be gained through this attainment of freedom, it is not a personal immortality, but one that is shared with all other rational beings. It is no more than an awareness or intuitive knowledge of the eternal truths and as such, it is, arguably, identical in all minds in that same intuitive state. Far from extending the individual self, Self-identification would seem to involve a loss, not only of our individual agency, but also of individual selves as they are dissolved or absorbed into the whole. Aware of the issue, Naess denies it: '[t]he widening and deepening of the individual selves *somehow* never makes them into one "mass"' (Naess 1989, 173), but Naess's own italicization shows that he is unable to account for how individuality is retained.

A further problem stemming from the high value Spinoza puts on knowing things *sub specie aeternitatis* is that in so doing, Spinoza appears to favour the intellectual realm of eternal truths over the sensory world of nature that for Spinoza was accessible only through the false inadequate ideas of the first kind of knowledge. Plumwood (2002) considered the pursuit of an absolute eternal perspective – the 'view from nowhere' – as especially damaging in ecological terms, for in privileging the objective viewpoint, the finite, limited and relativistic perspectives from which we actually engage with the natural world are correspondingly devalued, along with a similar devaluing of the sensed physical world itself. In its promotion of intellectually known eternal truths and corresponding belittling of sense experience, Spinoza's system of philosophy harks back to what Plumwood (2002, 46–47) identified as a Platonist or Greek model in which the knower (mind) and the known (the eternal truths) are equally and highly valued, but in which the 'sensory and material world' is discarded as 'unworthy' on the ground that it cannot yield true knowledge. Yet deep ecologists inspired by Spinoza do care a great deal for this physical and sense-perceived world that we perceive so inadequately in our 'random experience' (*Ethics* II, P40 Scholium 2: Gebhardt II 122; Curley 477) and that Naess himself celebrated for its therapeutic diversity of colours, sounds, tastes and other sensations and in whose immersion we are brought back to the 'good' (Naess 1969, 63).[43]

Evidently, Spinoza's philosophical system requires revision if it is to advocate the ecological value of the world as perceived through the senses. Any modifications would need to find a way to promote the value of sense experience without undermining the absolute monism and high estimation of the intellectual love of God or Nature on which Ecosophy T's Self-realization! norm so strongly depends. It is not, however, my intention to embark on such a task here. Instead, let us take our lead from Naess's own admission that 'one size does not fit all' and that the same destination may be reached from any number of different ecosophical paths. Alongside Descartes and Spinoza, Leibniz completes the canonical trio of early modern rationalists.[44] In light of the ecological difficulties associated with Cartesian dualism and in recognition of the problems arising from Spinoza's absolute monism, it is fitting now to consider the ecological potential of the philosophy espoused by the third and youngest philosopher in our rationalist canon: the optimistic, nominalist, relational and dynamic *pluralist* metaphysics of Leibniz.[45]

Notes

1 Paul Crutzen and Eugene Stoermer argue in favour of dating the start of the Anthropocene in the last quarter of the eighteenth century (Crutzen and Stoermer 2000).
2 The classic critique is Merchant 1989 [1980].
3 See, for instance, Catherine Ward Thompson et al. (2011, 2012).
4 Gaukroger (1995) and Clarke (2003) offer more Hobbesian readings of Descartes on this topic, suggesting that Descartes' animals do have sensations that are more akin to our own, but differing from ours in that the animals lack conscious awareness. In other words, nonhuman animals do not know that they are having experiences.
5 Lloyd (1993 [1984]). Carolyn Merchant (1989 [1980]) too assigned dualism a key role in the story of how humanity lost touch with the rest of the natural world. Others have questioned Descartes' anti-environmental credentials. Cecilia Wee offers a spirited defence of Descartes, even suggesting that he be regarded as the 'forerunner' of modern ecocentrism (Wee 2001, 276). Charles Taliaferro has also questioned Descartes' anti-environmental image (Taliaferro 2001, 36–37).
6 For instance, Simon Hailwood theorizes that the very 'otherness' of the natural world is what gives it intrinsic, and not merely anthropocentric instrumental value (Hailwood 2000).
7 Along with many others of his age, Descartes considered vivisection, even on dogs, an acceptable tool in the pursuit of scientific knowledge (*Description of the Human Body*, Pt 2: AT XI 243; CSM I 318).
8 Locke did amend the Cartesian view by including solidity in the definition of bodies as extended things. This opened up the possibility of empty space: space is extended, but bodies are both extended and solid. See *Essay* 2.4. For Locke's opinion on animals' experiences, see *Essay* 2.10.10.
9 Yolton 1984 and 1991.
10 For example, evolutionary biologist, Sam Brown, argues that some bacteria (for instance, the common bacterium, *Pseudomonas aeruginosa*) are sensing, experiencing entities that communicate, cooperate and compete with each other and with members of other species in social environments (e.g. Brown and Johnstone 2001; Cornforth et al. 2014). Neurobiologist and biochemist Jonathan Delafield-Butt has detected goal-directed movement in the unicellular *Paramecium caudatum* and interprets the experimental evidence in accordance with general tau theory as indicating that 'single-celled *Paramecium caudatum* guides their locomotion to an electrical goal using prospective sensory and intrinsic information' (Delafield-Butt et al. 2012, 284). For a Whiteheadian interpretation of perceptuomotor teleological or goal-directed movement in larger animals, applicable in principle also to smaller organisms, see Delafield-Butt 2014. The feelings and emotions of larger mammals are also discussed in Packard and Delafield-Butt 2014. Meanwhile, in physics, Carlo Rovelli and Matteo Smerlak postulate that in relational quantum mechanics, 'an observer can be *any* physical system' (Smerlak and Rovelli 2007, 429) and sound remarkably Leibnizian as they claim that even though it is not conscious, '[a]n atom interacting with another atom can be considered an observer' (Smerlak and Rovelli 2007, 430n9). It should be noted too that in philosophy, Galen Strawson advocates a 'new materialism' that regards features more commonly attributed to the mind – the mental – as *irreducible* properties of matter itself (Strawson 2006).
11 Sprigge 1971, 166–168; 1980, 41–44; Nagel 1974.
12 I use 'experience' here, as elsewhere, sense explained in note 12 of the Introduction (p. 9).
13 Had Hobbesian materialism triumphed over Cartesianism, the physical sciences might have developed in a quite different direction. Hobbes's notion of living matter does not reduce sensations, feelings and thoughts to the movements of inanimate

particles of moving matter. Consequently, under Hobbes, neither psychical experience nor physical motion is prioritized. Hobbesian human beings, for instance, are not just bodies that move; they are also sensing, rational living bodies. See Frost 2005, esp. 501–503, 516n22.

14 For discussion of Whitehead, see chapter 3, pp. 59–60. See also Introduction, pp. 3–4.

15 On Spinoza's instructions, the *Ethics* was published posthumously. Accordingly, it appeared in the *Opera Posthuma* in 1677, the year of his death. A Dutch translation, probably from an earlier version, also appeared in 1677, in the *De Nagelate Schriften*. The critical edition is that of Gebhardt (Spinoza 1925). The translation used here is that of Edwin Curley (Spinoza 1985).

16 Definitions that are real, as opposed to merely nominally agreed, are taken as being 'true' definitions. Postulates, as the word suggests, are reasonable assumptions that have not been formally proven to be true.

17 'By substance', wrote Spinoza, 'I understand what is in itself and is conceived through itself, i.e., that whose concept does not require the concept of another thing, from which it must be formed' (*Ethics* I, D3: Gebhardt II 45; Curley 408). A useful study of the concept of substance in the seventeenth century is Woolhouse 1993.

18 Here Spinoza refers to 'God, *or* Nature' (Gebhardt II 206; Curley 544). The idea is found in earlier writings too, as in the *Short Treatise on God, Man, and His Well-Being:* 'Nature consists of infinite attributes, of which each is perfect in its kind. This agrees perfectly with the definition one gives of God' (Gebhardt I 22; Curley 68).

19 'By God I understand a being absolutely infinite, i.e., a substance consisting of an infinity of attributes, of which each one expresses an eternal and infinite essence' (*Ethics* I, D6: Gebhardt II 45; Curley 409).

20 The question whether Spinoza's God possesses infinitely many or finitely many attributes is addressed by, among others, Haserot (1972) and Bennett (1984, 75–79).

21 See also *Ethics* I, P29 Scholium: Gebhardt II 71; Curley 434.

22 In the *Short Treatise*, Spinoza distinguishes universal and particular senses of *natura naturata*. The former modes depend on God immediately; the latter depend on God indirectly through the universal modes (Gebhardt I 47; Curley 91).

23 *Ethics* II, P13.

24 The common opinion follows Della Rocca's understanding that even rocks and hammers are 'in some sense, animate and possess mental states' and furthermore, that 'Spinoza has no principled basis on which to claim that not all mental states are conscious ones' (Della Rocca 1996, 9). However, on Wolfson's reading of Spinoza, the lower organisms are neither living nor conscious (Wolfson 1983 [1934], II 58–59).

25 Spinoza graphically illustrated this conception of the world as a whole comprising part-wholes in his example of the worm in the blood in a letter to Henry Oldenburg (20 November 1665: Gebhardt IV 171–173; SL 193–194).

26 See previous note, note 25.

27 Insofar as Spinoza understands falsity as a privation of truth, inadequate ideas are false because they fail to contain the reasons and logical connections that demonstrate fully their necessity and truth. Adequate ideas, by contrast, contain all the internal marks of true ideas: 'By an adequate idea, I understand an idea which, insofar as it is considered in itself, without relation to an object, has all the properties, or intrinsic denominations of a true idea' (*Ethics* II, D4: Gebhardt II 85; Curley 447).

28 See, for example, Deleuze 1988.

29 The lecture is summarized in Naess 1973.

30 Naess and Sessions 1984.

31 Taylor 1986 argues for essentially the same compromise in the face of conflicting needs of human and nonhuman.

32 'T' is a reference to 'Tvergastein', the cabin in Norway that Naess regarded as his spiritual home.

33 See, for instance, Ecce de Jonge's critique (de Jonge 2004).

34 On interconnectedness, see Naess 1977, 48.

35 In chapter 5, we will discuss the role of relations in the formation of individuals' identities in the context of Leibniz.

36 In chapter 6, we shall find that Leibniz too advances the view that acting in others' best interests and putting their interests above our own is actually the most effective way to serve our own best interests.

37 *Pace* Spinoza, however, Naess prefers not to emphasize that this perfect Nature is God (Naess 1969, 67).

38 As we have seen, Naess qualifies his nonanthropocentric ecocentrism with an 'in principle' clause that allows for 'vital' human needs to be given priority in situations of conflict. We address the conflict of interests and desires in our discussion of Leibniz's relational space in chapter 6. Here it is sufficient to note that both philosophers accept that the true good of any one individual is also a true good for the others and the whole.

39 More radical offshoots, such as Earth First!, have adopted more radical direct-action measures and accord less importance to the need to provide philosophical grounding for their actions.

40 In sharp contrast to Leibniz, Spinoza held that bodies may be *either* in motion *or* at rest (*Ethics* II, P13 ScholiumA1).

41 For an opposing opinion, see Rice 1996.

42 Ironically, insofar as Spinoza's world is a manifestation of multiplicity within the unity of God it conforms to the Enlightenment rationalist criteria for beauty that derive, not from Spinoza, but from Leibniz.

43 Naess has been charged, with some justification, with caring only for 'wilderness' at the expense of the rest of Spinoza's *natura naturata*. Indeed, his personal support for a return to a simpler, less technological and more ecologically sustainable way of life, epitomized by his naming Ecosophy T after his Norwegian cabin, Tvergastein, cannot be supported by an absolute monism in which everything is included within the whole and is a modification of the one divine Substance and in which there is no basis for an ontological distinction between the mountain and the factory farm or between the mosquito and the genetically engineered mouse. The built environment is as much a part of the universe as a whole as is the mountain. Should not wide- or Self-identification incorporate the factory or high-rise flat as much as the mountain?

44 The standard classification of Descartes, Spinoza and Leibniz as rationalists and Locke, Berkeley and Hume as empiricists is not beyond criticism. Even if standard divisions were clear-cut, the rationalist canon has been unduly restricted. Others, such as Malebranche, deserve to be included. However, this is not the place to critique the standard classification or to justify Leibniz's place in the canon.

45 I am not alone in believing that a pluralist ontology may be worth pursuing. In his review of de Jonge (2004), Edward Butler remarks: 'Deep ecologists have erred in grasping at a global, totalizing "non-dualism" instead of working out a complex, *pluralistic* picture of the human in nature' (Butler 2005).

2   Leibniz

Il n'y a rien de si beau ni de si satisfaisant que d'avoir une veritable connoissance du systême de l'Univers, non seulement à l'égard des corps, mais encor à l'égard des substances en general, et sur tout à l'égard de la nature divine et de celle de nostre ame, et même des ames en general.

(Leibniz to Thomas Burnett, Hanover 17/27 July 1696: GP III 182)

The true philosophy ... must give us an entirely different concept of God's perfection, one that will be of use in both physics and ethics.

(Leibniz to Christian Philipp, January 1680: GP IV 284; L 273)

One of the last great polymaths, Leibniz (1646–1716) was trained in law and employed, first as secretary to a secret alchemical society in Nuremberg,[1] then, in the court of the Elector and Prince-Archbishop of Mainz, Johann Philipp von Schönborn, he served as assistant to court jurist Hermann Lasser, drafting proposals for a complete reform of the legal system. While still employed by Schönborn, Leibniz also acted for his patron and supporter, Baron Johann Christian von Boineburg, on a secret diplomatic mission to Paris and assumed the role of tutor to Boineburg's son, Philipp Wilhelm and Schönborn's nephew, Melchior Friedrich von Schönborn.[2] Leibniz remained in Paris even after Boineburg's death at the end of 1672, but in 1676 he took up residence in the court of the dukes of Brunswick-Lüneberg-Calenberg (Hanover), serving initially as court counsellor and librarian, then as Privy Counsellor under Duke Johann Friedrich. Under the second duke, Ernst August, Leibniz was assigned the role of court historian, with explicit instructions to research and write the history of the royal House of Brunswick. Under the third duke, Georg Ludwig, Leibniz played a background but nevertheless key role in negotiations that saw Georg Ludwig ascend the English throne as George I in 1714. However, employment details do no more than skim the surface of Leibniz's range of activities, for he was also a philosopher, mathematician, inventor, jurist, linguist, physicist and geologist. In philosophy, Leibniz is best known for his theory of monads, the doctrine of pre-established harmony, and his optimistic belief that this world is the best of all possible worlds. Mathematicians associate his name with the development of the differential calculus, in particular for his introduction of the integration symbol (∫) and what has come to be known as the Leibniz formula for π. Computer scientists recognize Leibniz as the inventor of binary arithmetic on which so much of modern-day digital computing relies and he himself invented a sophisticated calculating machine. In physics, Leibniz is applauded for his work in dynamics and in particular for his formulation of *vis viva* (living force) as $f = mv^2$, and for his theory of space and time as relative. Leibniz made important contributions in the life sciences too, not least through his development of a coherent conception of living beings as organisms.[3] The specific examples given here comprise only a fraction of the contribution Leibniz made to these and other fields of study that include geology, linguistics, formal logic, jurisprudence and library science, as well as newer disciplines such as psychology and sociology.[4]

Leibniz's early education had given him a firm grounding in the classics. His father, Friedrich Leubnitz, had been Professor of Moral Philosophy at the University of Leipzig. Friedrich died when Leibniz was only six years old, but from age eight, Leibniz was allowed unrestricted access to his father's library, which Leibniz used to supplement his formal schooling to the full. It was in his father's library, according to his own report, that he taught himself Greek and Latin and immersed himself in the works of the Greek and Roman philosophers.[5] Later, as a student at the University of Leipzig, Leibniz was taught by the Reformed Aristotelian Jacob Thomasius. Leibniz's own philosophy would later incorporate many features of Aristotle's thought, suitably adjusted, in line with the Reformed tradition, for compatibility with the seventeenth-century mechanical philosophy.[6] It was after he arrived in Paris, however, that Leibniz was properly introduced to the 'new' Cartesian philosophy. There he met with French Cartesian Nicolas Malebranche and Antoine Arnauld, author of the Fourth Set of Objections to Descartes' *Meditations on First Philosophy*. Leibniz also attended regular meetings of the Académie Royale des Sciences in the Royal Library and it was there that he met the mathematician, Christiaan Huygens, under whose instruction, Leibniz advanced his own mathematical skills.[7] Another key contact in Paris was Baron von Tschirnhaus, a friend of Spinoza. Tschirnhaus had in his possession a copy of the manuscript of Spinoza's as yet unpublished *Ethics* and possibly also a copy of Spinoza's *Short Treatise on God, Man, and His Well-Being*.[8] While he may not have allowed Leibniz to read either manuscript, Tschirnhaus certainly discussed Spinoza's philosophical ideas with the Hanoverian philosopher. Leibniz was so intrigued by what he had learnt that en route to Hanover from Paris, in addition to stopping over briefly in London, he insisted on visiting Spinoza at his home in the Hague. We possess no detailed account of what was discussed at the meeting of these two great figures,[9] but it is clear that Spinoza, the man himself and his philosophy, had a profound effect on the young Leibniz, although Leibniz quickly realized that Spinoza's monist necessitarianism was incompatible with human freedom and autonomy.

The London stopover was Leibniz's second trip to the capital. He had already met founding members of the newly formed Royal Society of London, the English virtuousi, including the corpuscularian mechanist Robert Boyle, during his first visit. On that occasion, Leibniz demonstrated his new calculating machine at a meeting of the Royal Society. Unfortunately, the machine did not perform to full expectation on the day, but if it had, it would have astounded the audience with its facility to perform not only addition and subtraction, but also multiplication and division. Leibniz was subsequently elected as a Fellow of the Royal Society and established regular correspondence with the Secretary Henry Oldenburg, through whom Leibniz was kept well informed of intellectual developments in England.

It was not until the late 1680s that Leibniz secured agreement from his employers to take a second European trip, planned first to southern Germany, then extended to take in Austria and Italy, in search of evidence of the ancestral links between the House of Brunswick-Lüneberg and the Italian House of Este.[10] Leibniz did find crucial evidence,[11] but he also spent a large portion of his time on more academic pursuits, on mathematics, physics and philosophy. Towards the end of his trip, he was fortunate to meet the Franciscan professor, Michel Angelo Fardella. Leibniz's record of this conversation is of interest for what it reveals on his thinking on corporeal substance (FC 317–323; AG 101–105).[12] The intellectual friendship and correspondence that followed lasted nearly twenty-five years.

Over the course of his life, Leibniz enjoyed many other intellectually engaging friendships, the deepest perhaps being his friendships with Sophie, wife of the duke Ernst August, and her daughter, Sophie Charlotte. When offered, he seized opportunities to engage in conversation with visitors to the Hanoverian court whose numbers included the English freethinker John Toland and the alchemist 'scholar-gypsy' Franciscus Mercurius van Helmont[13] and he maintained an extensive correspondence, through which he conducted significant philosophical and theological debates with, among others, the Jansenist priest and astute critic of Descartes, Antoine Arnauld; the Cartesian Burcher de Volder; the Jesuit priest Bartholomew Des Bosses; and priest and friend of Newton, Samuel Clarke. It is through these correspondences and Leibniz's numerous notes, as much as from his published work, that we gain entry to the mind of this truly 'universal genius'.[14]

Outline of Leibniz's philosophy

Leibniz's mature metaphysics is one of monads. These are simple indivisible unities. Our own minds are monads. The experiencing centres of other living beings – the souls of animals or the substantial forms or entelechies of plants and smaller creatures – are monads. The term 'entelechy' derives from the Greek terms *en* (within), *telos* (end, goal, perfection) and *ekhein* (to have). Hence, Leibniz understands 'entelechy' to signify an entity that has perfection insofar as it contains its end or goal within itself, that is, as a being that contains in embryonic form all its future states and which also possesses the means to bring these future states to fruition, for the entelechy is the permanent '*Act*' or primitive active force that 'carries with it not only a mere faculty for action, but also that which is called "force", "effort", "conatus", from which action itself must follow if nothing prevents it' (*Theodicy* §87: GP VI 150; H 170). In this way, the entelechy, as 'Act', is also 'a realization of potency' (ibid.). That is to say, it is a primitive active force that gives rise to the succession of transitory actions or appetitions (appetites, desires, volitions) by which the entelechy progresses from one perception or perceptual state to the next. Because each monad is an active substance possessing some degree of perfection, Leibniz acknowledges that the term 'entelechy' can be used in a generic way to apply to all monads (*Monadology* §18). He grants that the term 'soul' can be used in the same way on the ground that all created monads have some kind of perception and appetition, whether these be minds, souls or substantial forms. However, his preference is to reserve the term 'soul' for those monads that can remember their perceptions – for instance the self-conscious rational minds of humans or the conscious souls of the higher animals – and to employ 'entelechy' (or 'substantial form') in relation to those monads that lack memory (*Monadology* §19).[15]

The term 'monad' began to appear fairly regularly in Leibniz's writings in the 1690s.[16] Prior to this, Leibniz had spoken only of 'individual substances'. In his 1686 *Discourse on Metaphysics* and the correspondence with Arnauld that ensued, the examples of individual substances he offers are human historical figures, such as Alexander the Great and Julius Caesar, and the first man 'Adam'. Nevertheless, Leibniz's nominalism is applied consistently throughout the natural world: the only things that exist are particular individuals. Of these, there are infinitely many, constituting a biophysical plenum of living entities, ranging from the most spiritual and angelic, through humans, animals, fish and plants, to protozoa and beyond. The essences of all of these individual substances are known to the divine mind in the form of 'complete concepts'. The theory of complete concepts relies on Leibniz's commitment to a 'containment' theory of truth that holds that:

> all true predication has some basis in the nature of things and that, when a proposition is not an identity, that is, when the predicate is not explicitly contained in the subject, it must be contained in it virtually. That is what the philosophers call *in-esse*, when they say that the predicate is in the subject. Thus the subject term must always contain the predicate term, so that one who understands perfectly the notion of the subject would also know that the predicate belongs to it.
>
> (*Discourse on Metaphysics* §8: GP IV 433; AG 41)

Each individual substance has 'a notion so complete that it is sufficient to contain and to allow us to deduce from it all the predicates of the subject to which this notion is attributed' (*Discourse on Metaphysics* §8: GP IV 433; AG 41). As finite beings, we can never complete the deduction, but the infinite 'God, seeing Alexander's individual notion or haecceity, sees in it at the same time the basis and reason for all the predicates which can be said truly of him' (*Discourse on*

Metaphysics §8: GP IV 433; AG 41).[17] Thus God knows that at a specific point in his lifespan, Alexander will become King. He knows too that Caesar will choose to cross the Rubicon, or to take a nonhuman example, God knows that this particular bird will at a certain point in time damage its wing. Every detail of the bird's life can be deduced from its complete concept, everything that will ever be predicated of it during its life, from when and where it builds its nest, in what weather conditions, what food it finds, when it goes hungry and when it is able to feast. In short, Leibniz held that the details of each and every individual's entire life experiences are contained in, and deducible from, their unique and individuating complete concepts.

The terminological shift to 'monads' coincides with Leibniz's official announcement of his discovery that the true nature of substance rests in the concept of 'force'. The idea had been forming in Leibniz's mind since the late 1680s,[18] but Leibniz waited until 1694 to make a public announcement in a short but important article in the *Acta Eruditorum* of his discovery that,

> the concept of *forces* or *powers*, which the Germans call *Kraft* and the French *la force*, and for whose explanation I have set up a distinct science of *dynamics*, brings the strongest light to bear upon our understanding of the true concept of *substance*.
>
> (*On the Correction of Metaphysics and the Concept of Substance*: GP IV 469; L 433)

In this paper, Leibniz is concerned with his science of dynamics and the physical forces of corporeal substances. However, while his new conception of substance provides insight into the nature of bodies as forces, he also claims that 'there follow from it primary truths, even about God and minds' (GP IV 469; L 433). God is pure active force, but only God is entirely active; minds or souls and their close relations, substantial forms or entelechies, are primitive active forces, but are always also limited by some primitive passive force, which Leibniz conceives as 'primary matter'. Nevertheless, no created monad is ever entirely passive either; all have some primitive active force.[19] It is therefore around this time that Leibniz starts to describe the monad as a conjunction of primitive active and primitive passive forces: created monads are combinations of metaphysically fundamental active and passive primitive forces.[20]

Active force brings with it the ability to change over time. That the states of monads continually change from moment to moment is one of their most basic features (*Monadology* §10). Even in the 1686 *Discourse on Metaphysics*, a certain temporality of past, present and future is built into the individual substance's complete concept. The predicates are related in such a way that when created the substance's experiences unfold in temporally sequenced order. Nevertheless, Catherine Wilson is right to point out that Leibniz's notion of the individual substance as the logical subject from which predicates are deducible presents an essentially 'static picture' of the individual substance (Wilson 1989, 160). By the early 1690s, when the monad as a primitive force has come to the fore, the emphasis has come to focus on the development and changes that the substance undergoes as its essence unfolds.[21] With the introduction of force as the essence of substance, Leibniz's notion of the complete concept of the substance is transformed into the notion of a 'rule' or 'law' that generates the sequence of ordered predicates attributable to that particular substance.[22] When created the monad's primitive active force unfolds its sequence of experiences (its perceptions or perceptual states) in accordance with this law; its appetitions (the actions or modifications of its primitive active force or 'internal principle') take the monad forward from one perception to the next (*Monadology* §15: GP VI 609; AG 215).[23]

Leibniz conceived this force as signalling in the substance an active internal spontaneity, contrasting it with the active power postulated by the scholastics. The latter is merely a potentiality to act that requires some external stimulation before actual activity occurs in the substance. As we have already noted, however, Leibniz's substances always act. Likening them to Aristotle's '*first entelechies*', he declares that the primitive forces 'contain not only *actuality*, or the mere fulfilment of a possibility, but also an originating *activity*' (*New System*: GP IV 479; WF 12). All the same, while no created monad is ever without some primitive active force, this active force is always combined with primitive passive force or primary matter. This passive force limits the monad's activity and prevents it from perceiving or representing the world with full clarity and distinctness.[24] The monads' appetitions do not always achieve the fullness of the perception that would result in their distinct and adequate perception of the whole world (*Monadology* §15). When they do, however, the monad becomes conscious of itself, aware of its own pleasure and pain and capable of making rational choices about its future courses of action. In these cases, the monad's appetitions are rational appetites or volitions. When the appetitions give rise only to conscious sensations, the monad's appetitions are desires. In some cases, a person may be sufficiently self-aware to know he has the desire without knowing *why* it is desirable; in other cases, he may be aware only of the desire itself, without being conscious of it as his own. Animals that have consciousness but lack self-consciousness might be supposed to have desires of this sort. Most appetitions, however, are so slight or imperceptible that they consist only of vague stirrings of disquiet. In all cases, though, appetitions are 'impulses' that propel the monad from one insensible or unconscious perception to the next (*New Essays*: A VI vi 192; RB 192).[25]

Insensible or unconscious perceptions are states typical of the simple or 'bare' monads (*Monadology* §24) that have yet to begin their progression toward perfection (*On the Ultimate Origination of Things*: GP VII 308; AG 155). They are found in other monads too when they fall into a faint or dreamless sleep (*Monadology* §20). At these times, our perceptions are so obscure that we are not at all aware of them. Although we may have self-conscious distinct thoughts when awake, our sense perceptions of bodies and their sensible qualities are always confused. When a perception of an object is confused, it is sufficiently clear for us to be able to distinguish its parts or individuating features.

Thus, we see as green what is really a mass of blue and yellow particles (*Meditations on Knowledge, Truth, and Ideas*: GP IV 426; AG 27). We feel the smoothness of the water-washed stone, but only vaguely and insensibly perceive its atomic and subatomic constituents. Our sense organs do not give us distinct perceptions of the infinitesimally small parts of matter. Microscopes can take us a little way in, but more distinct perceptions of the constitution of bodies require input from reason and intellect.

Monads are not only perceiving, appetitive beings; every monad also has an organic body:

> organic bodies are never without souls, and ... souls are never separated from organic bodies ... Thus I do not at all recognize entirely separated souls in the natural order or created spirits entirely detached from any body ... creatures free or freed from matter would at the same time be divorced from the universal bond, like deserters from the general order.
>
> (*Considerations on Vital Principles and Plastic Natures*: GP VI 545–546; L 590)[26]

Organic bodies are aggregates of substances, each unified by the monad that dominates it. Through its body, the monad gathers information about external things (that is, it perceives the world) and also attempts to fulfil its appetites and desires.[27] With its organic body, each monad exists as a living creature, a soul–body unity that Leibniz calls a 'corporeal substance' (to de Volder, 20 June 1703: LV 264–265).[28] Thus, for Leibniz, each finite substance is a living, embodied being. The model is drawn from his experience of himself and then imaginatively extended to all living things. Leibniz experiences himself, not as a disembodied 'I', but rather as an 'I' in a body. Drawing on the Principle of Uniformity, he infers from his own embodied experiences that other living things are not merely mechanical organic bodies, but that they too are experiencing, perceiving beings.[29] Not all monads are self-reflective, of course. Not all are aware of themselves as embodied 'I's. However, this does not mean that they do not experience the world from their own embodied perspectives. They do. All monads experience the world through the lenses of their bodies. Each one represents the whole of the universe by exactly mirroring, although for the most part insensibly and confusedly, the changes that the bodies of others effect on their own.[30]

All bodies are spatially extended composite things. They must, Leibniz maintains, be composites of simple indivisible monads (*Monadology* §§1, 2). However, Leibniz also thinks that bodies, as extended, are actually divided into increasingly smaller extended parts that are also, as extended, further divided, ad infinitum. We do not find monads at the end of a never-ending infinite series. Instead, we find them at the every stage on the way down. A human mind is a monad that unifies the substances that make up its human body. Together, human-mind and human-body, comprise a complete human being. The soul of a dog is a monad that unifies the substances that make up its dog-body.

Dog-soul and dog-body together make up the complete dog. As for the sub-stances in these human and dog bodies, they too are monads with organic bodies. For instance, each cell-entelechy unifies the substances that make up its cell-body and with them comprise the complete living cell. The same principle holds for all the parts of the cell and for the parts of the parts, and so on ad infinitum. There are simple monads 'all the way down', each unifying its own organic body, which in turn ensures that there are also monad–body unities – indivisible corporeal substances[31] or living creatures – 'all the way down'. As he explains in the *Monadology*:

> each living body has a dominant entelechy, which in the animal is the soul; but the limbs of this living body are full of other living beings, plants, animals, each of which also has its entelechy, or its dominant soul.
>
> (*Monadology* §70: GP VI 619; AG 222)

Leibniz believes that the divisions of matter proceed to infinity and thus, 'there is a world of creatures, of living beings, of animals, of entelechies, of souls in the least part of matter' (§66: GP VI 618; AG 222). With animated living beings at every stage of the division, the universe contains nothing that is 'fallow, sterile, or dead' (§69: GP VI 618; AG 222). Matter, he surmised, was like 'a garden full of plants' or 'a pond full of fish' in which 'each branch of a plant, each limb of an animal, each drop of its humors, is still another such garden or pond' (§67: GP VI 618; AG 222). There are no gaps in nature.[32] The Principles of Sufficient Reason, of the Identity of Indiscernibles, and of Perfection preclude a vacuum.[33] In the plenum, even the water, earth and air between the fish and plants are filled with living creatures, too minute for us to discern with the naked eye (*Monadology* §68: GP VI 618; AG 222).

Filled with animate bodies, matter is everywhere in constant motion. Nothing remains the same from one moment to the next. The water, air and earth between the fish and plants abound with life and movement, replete with living, experiencing and feeling creatures. Even inanimate objects, like tables and chairs, are composed of living creatures that never remain still even for an instant, the relations between the constitutive parts in constant flux. There is also constant interchange of parts across body boundaries. The boundaries of bodies are fluid, not fixed, as parts move from one body to another, from environment to self and from self to environment:

> we must not imagine ... that each soul has a mass or portion of matter of its own, always proper to or allotted by it ... For all bodies are in perpetual flux, like rivers, and parts enter into them and depart from them continually.
>
> (*Monadology* §71: GP VI 619; AG 222)

Motion in bodies indicates the presence of derivative active force. This, Leibniz reasoned, is the physical modification of the primitive active forces in the monads. Similarly, the resistance and impenetrability of bodies signals the presence of derivative passive force, the physical modification of the monads' primitive passive force. Derivative forces, deriving from the metaphysically primitive forces, constitute the essence of body or matter. Leibniz's materially extended bodies are essentially coalescences of forces embedded within forces. Whereas Descartes had conceived the notion of extension as metaphysically primitive and not further analysable, Leibniz insisted that the concept of extension is analysable into the notions of plurality, continuity and coexistence. Each extended body forms one continuous mass through the repetition or diffusion of the plurality of coexisting derivative forces that comprise the bodies of the living creatures within it and which are in turn the modification of its dominant monad's primitive force.[34]

Thus, the notion of force lies at the heart of Leibniz's rejection of the Cartesian account of passive matter as simple extendedness in length, breadth and depth. Matter is not merely passive. It is both active and passive.[35] A moving body, activated by its internal derivative active force, also resists the actions of other bodies upon it by its own derivative passive force. All is in harmony: the activity and passivity of one body perfectly accommodated to the activity and passivity in the others. Nevertheless, bodies move and resist only by their own forces. They are not pushed into action by external bodies. Their activity is limited by their own internal passivity, although the amount of this passivity at any one moment is determined by the need to balance the passivity of one body with the activity in others. Hence,

> one created substance receives from another created substance, not the force of acting itself, but only the limits and the determination of its own pre-existent striving or power of action.
>
> (*On the Correction of Metaphysics*: GP IV 470; L 433)

There is therefore no real causal interaction between bodies. There is no influx of force from one body to another. Each body acts and resists others only through its own forces, although exactly synchronized with the movement and resistance of others around it, such that they appear to be actually interacting. Bodies have been divinely preformed so as to guarantee that, solely through the operation of the laws of nature themselves, the forces of motion and resistance in one body are perfectly balanced with the forces in others. Leibniz regarded it as wholly unnecessary to postulate that any force is actually transferred from one body to another.[36] It is enough to acknowledge that when one body collides with another, its active force increases to the exact degree that the active force in the other decreases. Still, we may continue legitimately to regard one sub-stance as the reason for what happens in another, in accordance with whatever method provides the simplest and most intelligible causal explanation. It is, for example, simpler and more intelligible to attribute force and motion to the ship rather than to the water, even though,

> in metaphysical precision one is no more correct in saying that the vessel pushes the water to make this great quantity of circles which serve the

purpose of filling the place of the vessel than in saying that the water is pushed to make all these circles, and that it pushes the vessel to move accordingly.

(Draft of a letter to Arnauld, 28 November/8 December 1686:
GP II 69–70; LA 85)[37]

Just as there is no actual interaction between bodies, nor is there any between the mind or soul and its organic body, despite the body providing the perspective from which each soul perceives the world.[38] Nevertheless, when our eyes and brains are altered in response to the light from the sun and when our bodies are affected physically by the physical changes in the environment as the clouds move across the sky and the effects of their motion ripple through the material world, there is also in the soul a perception of the sun and clouds in the sky that corresponds exactly to the physical state of our own bodies at that time.[39] There is harmonious agreement between the soul and its body, for the soul, following its own laws, constantly reflects the ever-changing relations between the parts of its body as they move and resist one another in accordance with their laws:

> the organized mass in which the point of view of the soul lies is more immediately expressed by it, and is in turn ready, just when the soul desires it, to act of itself according to the laws of the bodily mechanism, without either one interfering with the laws of the other, the animal spirits and the blood having exactly at the right moment the motions which correspond to the passions and perceptions of the soul.
>
> (*New System*: GP IV 484; WF 18)

It is no surprise that they do, for the harmonious parallelism pre-established between them is grounded in the monads' essences or natures. Ultimately both the monads' appetitions and perceptions and the derivative active and passive forces (manifesting as motion and resistance) in their bodies stem from the same source: the primitive active and passive forces of the dominant monad or soul. The essence of a Leibnizian monad unfolds as both psychical and physical modifications in much the same manner as the essence of Spinoza's God or Nature gives rise to the dual series of idea-modes and body-modes. Leibniz's monads are like myriad finite versions of Spinoza's God. Despite their differences, for both Spinoza and Leibniz, the essence of a substance produces two kinds of modifications: ideas (Spinoza) or perceptions/appetitions (Leibniz) on the one hand, and, on the other hand, bodies (Spinoza) or bodies' derivative forces (Leibniz).[40]

Hence, Spinoza's conception of God as a substance that is both thinking and extended provides the model for Leibniz's embodied monads, the corporeal substances. But what Spinoza conceived within the context of an absolute monism, Leibniz echoes in a pluralist metaphysics. Where Spinoza admitted only one infinite Substance (God or Nature), everything else being mere modes of the divine attributes, Leibniz envisaged a plurality of embodied monads, each a substance in its own right, distinct from God, each possessing its own active force from which its perceptions and the motions of its body spontaneously flow. Leibniz's universe, as we have seen, is full of corporeal substances. It is a world replete with perceiving, appetitive dominant monads, each with its own organic body that is in turn also an aggregate of smaller corporeal substances. These subordinate corporeal substances are essential if the monad's primitive force is to be modified as the derivative force of its organic body: physical derivative forces can only be realized through the plurality of subordinate substances that comprise each monad's aggregate organic body.[41]

In respect of perceptions too, Spinoza's conception of God as a substance whose idea of himself is an idea of the whole world (*natura naturata*) is played out in a pluralist setting. Leibniz's monads are not idea-modes of the one single God-Substance, but this does not mean that they cannot, in some sense, hold the entire universe in their perceptual content. Within each finite individual substance, a version of the universe is contained. In a very real sense, our experiences are experiences of the whole in the one. 'Multiplicity-in-unity' is a characteristic feature of the Leibnizian world. In the pluralist setting, of course, there are infinitely many 'multiplicity-in-unity' experiences, each one representing the whole from its own unique and individual point of view. In this way, as we shall see in more detail in chapter 5, the Leibnizian monad is a relational being, standing in both internal and external relations to others. When John watches the blackbird on the fence, it is possible to distinguish two perceptions: the one, 'John-watching-blackbird' and the other, 'blackbird-watching-John'. Similarly, when the hawk watches the mouse, the mouse also perceives the hawk. Even if the mouse only subconsciously perceives the hawk, the same reciprocity of perceptions occurs: the hawk-watching-mouse perception and a corresponding mouse-watching-hawk perception. However, actual knowledge of this reciprocity is available only to self-conscious monads or minds, for in recognizing themselves as experiencing 'I's distinct from the things that they experience, they are able to form a conception of others as experiencers of the same world, but who experience that world from their own perspectives or points of view. In this, they come to recognize the other *both* as a distinct and complete entity in its own right, experiencing the world in its own way, *and* as an 'other' that is integral to its own being and from whom it cannot be separated (John would not be John did his perceptual state not include his perception of the blackbird).

In formulating the idea of each monad internally representing the same world from its own unique perspective, Leibniz drew upon a traditional conception of the human as a *mikros kosmos* or 'little world' structurally replicating in miniature the *makros kosmos*, the universe at large.[42] Leibniz was fond of the metaphor, invoking it at various points throughout his life:

> every substance is like a complete world and like a mirror of God or of the whole universe, which each one expresses in its own way, somewhat as the

same city is variously represented depending upon the different positions from which it is viewed. Thus the universe is in some way multiplied as many times as there are substances, and the glory of God is likewise multiplied by as many entirely different representations of his work.

(*Discourse on Metaphysics* §9: GP IV 434; AG 42)

Just as the same city viewed from different directions appears entirely different and, as it were, multiplied perspectively, in just the same way it happens that, because of the infinite multitude of simple substances, there are, as it were, just as many different universes, which are, nevertheless, only perspectives on a single one, corresponding to the different points of view of each monad.

(*Monadology* §57: GP IV 616; AG 220)

The mutual expression or representation of all by all is a harmonious ordering pre-established in the realm of possibility and requiring only the divine act of creation to begin the process of temporal unfolding in actuality. Just as there is an exact accommodation of bodies with each other and a parallelism of souls or entelechies with their bodies, so too, there is a precise isomorphic correspondence between monads. Each monad's perceptions arise spontaneously and naturally from its essence or law, without the need for prompting from any other monad. Nevertheless,

the perceptions or expressions of all substances mutually correspond in such a way that each one, carefully following certain reasons or laws it has observed, coincides with others doing the same – in the same way that several people who have agreed to meet in some place at some specified time can really do this if they so desire.

(*Discourse on Metaphysics* §14: GP IV 439: AG 47)

The absence of causal interaction does not signal disconnectedness. On the contrary, there is a thoroughgoing interconnectedness of all living things, built into the very essence of each individual being (chapter 5). The very construction of the law or concept of each individual is inextricably bound up with the formation of the essences of all the other individuals to which it relates and together with which it makes one of many possible worlds. On creation of the actual world, each individual's essence simply unfolds, spontaneously and independently, but in a harmonious conjunction with the others that was pre-established when this world was only one of many possible worlds.

For Spinoza, the world that exists is logically necessary, following automatically from God's unchangeable essence. Anything that does not exist can in principle be shown to be logically impossible or contradictory. Thus, for Spinoza, the actual is coextensive with the possible. Leibniz, in contrast, insisted that there are logical possibilities that are never actualized. There are many logically consistent possible worlds, but only one actual world. Logical possibility, therefore, does not, for Leibniz, entail existence. However, if Leibniz's God did not create this world purely on the basis of logical necessity, what reason is there for God to create this world rather than any of the others? In the absence of logical necessity, there has to be some other reason why God created this world rather than any of the other possible worlds: a sufficient reason that depends not on logical necessity, but on principles of perfection and goodness. Accordingly, Leibniz believes that God created this world not because he was absolutely necessitated to do so, but because God willed to create the most perfect or best possible world.

Thus is teleology – so conspicuously absent from Spinoza's thought – at the very heart of Leibniz's philosophy. God's *telos* or end was the creation of the best possible world. God acts in accordance with final causes, freely willing to bring into creation that which is overall the best, where the 'best' is defined in terms of perfection, goodness, variety, order and harmony. Teleological principles operate in each of the constituent parts of creation too. All living creatures strive to attain that which they perceive, albeit not always accurately, as the best. Souls and entelechies are governed by teleology, following the laws of final causation, taking into consideration criteria pertaining to the goodness and perfection of the anticipated outcome. The succession of perceptions is governed by perceptions of the good or the apparent good. Perceptions of the true goods, perceptions of the perfect varied orderliness in things and in the world as a whole – that is to say, the perception of the perfection or beauty of the world and its constituents – initiates feelings of pleasure and ultimately of joy and happiness in the perceiver, as well as feelings of love towards the beautiful, both in created things and in their Creator (chapter 6). Leibniz's philosophy also leaves open the optimistic possibility of a general progression of both individual and world towards even greater perfection.[43] Numerically, the order and variety of the world remains constant. So too, the overall degrees of monadic primitive and physical derivative force are conserved. However, the distribution of primitive force across the monads is compatible with the development of moral and aesthetic sensibilities (chapter 8) and the alignment of volitions with the will of God that arises from the knowledge that the true good – that which God wills – does not require that some benefit at the expense of others. On the contrary, it requires that all be allowed to flourish.

Since every soul or entelechy operating according to the laws of final causation is always attached to an organic body operating according to the laws of efficient causation, it follows that perception in the soul or entelechy is always accompanied by activity in the body. They stem, as we have seen, from the same primitive active force in the monad. It also follows that every event or state of affairs is explicable both in terms of final causes and in terms of efficient causes. Any situation can be described completely in terms of efficient causes or in terms of final causes (*Monadology* §36). A shepherd searching for his sheep in the hills does so because night is falling and he does not want his sheep to be in danger after dark (final causation) *and* because the muscles in his body expand and contract in appropriate ways to make his body move across the landscape (efficient causation).

In this way, both mechanism and teleology operate throughout the natural world as Leibniz conceived it. All the same, he insists that it is not appropriate to mix the two types of explanation (*Monadology* §17). It is not acceptable, for instance, to explain the motion of the shepherd's body by appealing to his desire to find his sheep nor is it appropriate to explain his changing perceptions of the landscape by appealing to the movement of the muscles in his body or the firing of the neurons in his brain. The physical state of the brain does not explain the soul's perception, nor does the soul's desire explain the movement of the body. For this reason, the natural sciences should, as far as possible, explain phenomena through the laws of motion and the physical qualities of bodies. Teleological explanations by final causation are more naturally at home in philosophical and theological contexts.[44] In the present context, we may say that ecology as a science of the interconnections between biological organisms and between them and their environments ought not to be confused with ecophilosophy as a philosophical discipline that introduces value judgements on human engagement with these organisms and environment. Both are important, but their methods and explanatory frameworks are quite different. Nevertheless, they inform one another and on occasion ecophilosophical insights may prove valuable in directing scientific ecological research in promising directions.[45]

However, when it comes to the question of why the laws that govern the motions of bodies are as they are, teleological explanation can, and must, be brought into play. In short, the laws of mechanics are justified on teleological grounds. Ultimately, Leibniz holds, efficient causation is subordinate to final causation. There is no mechanical explanation for the laws of motion themselves. Final causation must here be invoked. The reason why these laws obtain rather than others is because these are the laws by which bodies operate in the best possible world. The laws themselves were chosen because they form part of the envisaged end: the best possible world. In choosing which of an infinite number of possible worlds to create, God compared worlds in terms of their relative perfection and goodness, freely choosing to create the best of all the possible worlds. The laws of motion and other mechanical principles were not chosen in isolation from general considerations about the perfection of the entire created world. As ever, the overall determining factor is the consideration of the perfection or goodness of the end result.[46]

Notes

1 Antognazza 2009, 80.
2 Antognazza 2009, 145.
3 To fully appreciate the depth and range of Leibniz's contributions to the life sciences, I refer the reader to Smith 2011.
4 Appropriately, Leibniz was the founder and first president of the Society of Sciences of the Elector of Brandenburg. The Society was founded under the auspices of the

electoe, Frederick III of Brandenburg and renamed in 1701 as the Prussian Academy. Today it exists as the Berlin-Brandenburg Academy of Sciences and Humanities.

5 Antognazza 2009, 34–35. Antognazza rightly questions the veracity of Leibniz's reconstruction of his philosophical development from the ancients through the scholastics to the moderns, but there is no doubt that Leibniz was remarkably well-read.

6 Antognazza 2009, 54–57. See also Mercer 2001, 119–129, 168–169.

7 Antognazza 2009, 142–143.

8 See Lærke 2011, esp. 36.

9 Leibniz did however record that he had shown Spinoza his proof 'That a Most Perfect Being Exists' (18–21? November 1676: DSR 102–103).

10 Antognazza 2009, 281.

11 Antognazza 2009, 289.

12 For discussion, see Garber 2004 and Phemister 2005, 91–99.

13 Merchant 1979, 171.

14 Broad 1979 [1975], 3.

15 This is the convention that will be used throughout the following chapters.

16 For more detail on the use of the term, see Antognazza 2009, 352. Also see Rutherford 1995, 172n67.

17 The most comprehensive study of Leibniz on individual substances is di Bella 2005. See also Cover and O'Leary-Hawthorne 1999.

18 Wilson 1989, 156.

19 There is thus nothing in the world that is purely inanimate or without life. Even inanimate objects, such as tables and chairs, are, for Leibniz, aggregates of substances that have primitive active force. Pure passivity, in Leibniz's view, is an incomplete abstraction that cannot exist in reality (Leibniz 1854, 30).

20 See, for instance, Leibniz's letter to de Volder, 20 June 1703: LV 264–265. To avoid conflict with Leibniz's description of monads as soul-like primitive active forces alone, commentators have been tempted to conceive primitive passive force or primary matter as that which prevents the monad from attaining fully distinct perceptions and that explains why the monad's active force only reaches at best perceptions that are confused. In this way, the monad remains primarily a perceiving being, and the description of it as a soul or active force survives relatively unscathed (e.g. Rutherford 1995, 162–164). However, such readings tend to view primary matter's function as the principle underlying the resistance and extension of the monad's organic body as of secondary importance and thereby run the risk of undervaluing the physical world in favour of an idealist and more fundamental immaterial monadic reality of which the physical is a merely appearance. On my reading of Leibniz (Phemister 2005), the monad – or at least the monad as created – is an *embodied* perceiving being, whose '*primary matter* or mass [*molis*]' is associated with the 'passive force of resistance' in bodies (*On Nature Itself, or on the Inherent Force and Actions of Created Things*: GP IV 510; L 503). It is neither necessary nor appropriate to rehearse the arguments on either side of the debate here. For our purposes, we shall understand the monad's primary matter or primitive passive force as a real monadic force that explains *both* why monads have confused and insensible perceptions *and* why their organic bodies resist the actions of other bodies on them. This reading assures the reality and status of the physical world and is, as we shall see, not dissimilar to some of the notions of Leibnizian monads that were advanced in the eighteenth and nineteenth centuries.

21 Wilson 1989, 160.

22 See Nachtomy 2007, 70, 180.

23 Michel Fichant (2004) argues that there is rather less resemblance between the individual substances of Leibniz's *Discourse on Metaphysics* and the monads of his *Monadology* than my remarks about complete concepts and laws of the series might lead us to suppose. It is worth noting here, however, that even if monads per se are

quite different from individual substances, Leibniz's corporeal substances (that is, monads together with the organic bodies from which they are never separated, even in death) are just the kinds of things – human beings, sheep, worms and other living animal-like creatures – that he had previously regarded as individual substances.

24 Without passivity, there could be no plurality of substances, for each monad would be identical with the pure activity that is God (*Theodicy* §31: GP VI 121; H 142 and §64: GP VI 137; H 158).

25 See also Phemister 2005, ch. 10.

26 Leibniz often claimed that monads are always attached to organic bodies, but the point is especially prominent in his letters to Lady Masham, dated beginning of May and September 1704 (GP III 340; WF 205 and GP III 362–363; WF 219).

27 Technically, as we shall see below, the monad's appetitions effect changes in the monad's perceptual states and only indirectly through the union of the soul with its own body does it bring about changes to bodies in the wider world.

28 See also *On Nature Itself*: GP IV 511; L 503–504. The corporeal substance – the monad together with its organic body – draws a line of continuity between Leibniz's early 'individual substances' and his later monadology. See above, note 23 (p. 46–47).

29 E.g. Leibniz to Masham, beginning of May 1704 (GP III 339; WF 204). Also see *Considerations on Vital Principles and Plastic Natures* (GP VI 546; L 590). For discussion, see Phemister 2004, esp. 200–211. On the term 'experience' as inclusive of nonconscious states, see Introduction, note 12 (p. 9).

30 See, for instance, *Principles of Nature and of Grace* §3 (GP VI 598–599; AG 207).

31 See Leibniz's letter to de Volder, 20 June 1703: LV 264–265.

32 To de Volder, 3 April 1699: LV 68–71.

33 See e.g. Postscript to Leibniz's Fourth Letter to Clarke: GP VII 377–378; LC 43–45.

34 See Phemister 2005, ch. 3, esp. 57–69.

35 Leibniz explains in *On Nature Itself* that as well as impenetrability, 'there is a natural *inertia* opposed to motion in matter, so there is in body itself, and indeed, in every substance, a natural *constancy* opposed to *change*' (*On Nature Itself*: GP IV 511; L 503). However, (living) bodies can also initiate motion. Once in motion, each body also 'retains an impetus and remains constant in its speed … it has a tendency to persevere in the series of changes which it has once begun' (*On Nature Itself*: GP IV 511; L 503). From this he concludes that every corporeal substance has both extension and a '*primary entelechy* … a primitive motive force' and goes on to note that this is the soul or substantial form that, when combined with matter or primitive passive force, comprises the monad (*On Nature Itself*: GP IV 511; L 503–504).

36 In one sense of course, since bodies are in flux, forces are transferred from one body to another. The introduction of new substances and the expulsion of others – as for instance, in nutrition and excretion – are preformed mechanisms by which the modifications of the dominant monad's primitive force are realized in accordance with the dominant monad's essence, law or complete concept.

37 Consider, for instance, a dog gnawing at a bone. Under pre-established harmony, the bare-monads in the bone itself do initiate the movement of their own bodies, which bodies comprise the parts of the bone, but it makes more sense to attribute the greater active force, motion, distinct perception and effective appetition or desire to the dog rather than to the monads in the bone.

38 To Arnauld, 9 October 1687: GP II 113; LA 145.

39 Leibniz himself offers the example of being pricked by a pin – the physical damage to the body being registered as pain in the soul (to Arnauld, 9 October 1687: GP II 114; LA 146).

40 The Spinozist reading of Leibniz is developed at length in Phemister 2005 and rehearsed again in Arthur 2014.

41 For more detail, see Phemister 2005, ch. 8.

42 Conger 1922 briefly traces the fascinating history of the micro–macrocosmic theories from early Greek times to the twentieth century.

43 Like beauty, the notion of progress, so prominent in Leibniz's thought, is also absent from Spinoza's philosophy.

44 Leibniz does admit that teleological considerations can be helpful short-cuts in the natural sciences, especially anatomy, but also in optics. See *Discourse on Metaphysics* §22: GP IV 448; AG 54–55.

45 On these distinctions, see Naess 1989, 36–37.

46 From this, we may infer that the beautiful and ordered interconnections uncovered by ecologists ultimately require ecophilosophical explanation.

3 Leibniz's legacy

> The new monadology comes ... not to destroy but to fulfil the old.
>
> (Herbert Wildon Carr, *Cogitans Cogitata*, postscriptum:
> Carr 1930, 110)

At the time of his death, Leibniz and his often-controversial views were known throughout Europe. During his lifetime, Leibniz published only one monograph, the *Essays on Theodicy*; his preference was to present his ideas to the public in the form of journal articles. His *New System of the Nature and of the Communication of Substances*, in particular, published in the *Journal des Savants* in 1695, had elicited a wave of public responses from leading intellectuals of the day, including the prominent sceptics, Simon Foucher and Pierre Bayle.[1] Moreover, throughout his life, Leibniz had maintained a vast correspondence with an impressive array of the leading figures of the time. Consequently, his philosophical views are scattered across a range of papers, letters and personal notes, the full edition of which is still even today under way. Nevertheless, in the hands of Leibniz's one-time correspondent, Christian Wolff (1679–1754), a more condensed and systematic presentation of Leibniz's philosophy followed shortly after Leibniz died. Wolff's immensely popular lectures at the University of Halle presented a version of Leibniz's metaphysics as a deductively demonstrated system of philosophy in which philosophy is conceived as the 'science of the possible' and in which everything, even Leibniz's Principle of Sufficient Reason, is deduced logically from the Principle of Contradiction. Thus did Wolff set the scene for the development of a uniquely German style of philosophizing that Cassirer finds epistomized in Kant: 'his formulation of problems and systematic method' stemmed directly from the Leibniz–Wolffian philosophy, insofar as 'the philosophy of this era had clearly seen and recorded one of the great possibilities of the development of a uniform theoretical system of thought' (Cassirer 2009 [1951], 123). Leibniz's philosophy, with its emphasis on the notions of force, the dynamic activity of substances, multiplicity-in-unity, harmony, order and perfection, and its analysis of the way in which distinct perception of harmony connects with freedom, arouses feelings of pleasure, joy, happiness and love and leads ultimately to the rational appetite or will to virtuous action that is underpinned by knowledge of God and the true good, shaped 'all that the

German Enlightenment contributes in the fields of psychology, epistemology, ethics, aesthetics, and the philosophy of religion'. Always, Cassirer continues, 'science and systematic philosophy' in Germany 'found their way back to those fundamental questions first raised by Leibniz' (Cassirer 2009 [1951], 122–123).

Although Wolff eschewed using the term 'monad' and subordinated the teleological Principle of Sufficient Reason to the logical Principle of Contradiction (*Ontologia* §70: Wolff 1962), his Leibnizianism is evident in many other respects. On bodies, Wolff agreed that these are aggregates of simple substances (*Cosmologia Generalis* §176: Wolff 1964) and that the forces in bodies (Leibniz's derivative forces) derive from the forces of the simples (*Cosmologia Generalis* §196). On souls, Wolff agreed with Leibniz that the soul is an indivisible unity that is an active, spontaneous, independent and self-sufficient being, in which sensation and intellect are not sharply distinguished, but are instead different points on a continuum. Wolff agreed too that the soul's nature consists in the active representation of things, even though nothing can enter from outside (Cassirer 2009 [1951], 120).[2] Accordingly, Wolff also defended Leibniz's doctrine of pre-established harmony, both the harmony among the forces of bodies and the harmony of bodies and souls. Bodies possess both active motive force and passive inertial forces and they move and resist each other, not because of an influx of force from external bodies, but because of their own active and passive forces.[3] Nor is there any 'physical influx' from body to mind or from mind to body.[4]

Leibniz's doctrine of pre-established harmony was soon to be defended again, this time by Alexander Gottlieb Baumgarten (1714–1762). First a student and then a teacher at the University of Halle, Baumgarten published his *Metaphysica* in 1738/39, the first volume of his *Aesthetics* in 1750 and the second volume in 1758. Today, Baumgarten is most usually remembered as the 'father of modern aesthetics', a title that Frederick Beiser argues is in many ways more appropriate to Wolff (Beiser 2009, 48, 118). Beiser rightly awards Leibniz the status of 'grandfather of modern aesthetics' (Beiser 2009, 31) for it was Leibniz's conception of beauty in terms of the unity-in-variety that is the mark of perfection and the contemplation of which is pleasurable that was taken forward by Wolff and Baumgarten,[5] together with an emphasis on his Principle of Sufficient Reason, and the preservation of the link between the concepts of beauty and perfection and those of truth and morality, all of which laid the foundations for the tradition of German rationalist aesthetics that continued through Johann Christof Gottsched (1700–1766), Johann Joachim Winckelmann (1717–1768), Moses Mendelssohn (1729–1786), up to Gotthold Ephraim Lessing (1729–1781).[6] We shall return to Leibniz's ideas about beauty and perfection in the chapters that follow.

The notion of perfection and its cognates, beauty, truth and goodness, are characteristic of the optimistic metaphysics that Leibniz demonstrated with the aid of the Principles of Contradiction, Sufficient Reason, Continuity, Perfection and the Identity of Indiscernibles. Baumgarten's philosophical reasoning followed the same pattern, for his Leibnizian heritage is never far from the surface in his writings. Rejecting Spinozistic monism and necessitarianism, Baumgarten acknowledged possible worlds comprising possible individuals (*Metaphysics* §377: BM 170) and conceived existing finite created individuals or monads as simple indivisible, dynamic substances or forces (§396: BM 174).[7] All the same, Baumgarten did not just expound Leibniz's philosophy; he also adapted and extended it. For instance, he takes from Wolff the non-Leibnizian term, 'nexus', to refer to the relationality or connectivity of actual or possible things whereby one is the consequence or effect of another, its ground or reason or even ground and consequence of each other (*Metaphysics* §14: BM 102). In essence, things in relationship are nexuses. There are different kinds of nexus, many of which are in keeping with the various harmonies identified by Leibniz. A 'nexus of use', for instance, pertains when someone could or does use something for some purpose. So, we might say the fork and the diner constitute a 'nexus of use'. The universal harmony is also a nexus (§48: BM 109; §279: BM 151). There are also particular harmonies, all pre-established, as is the universal harmony of which God is the ground. The realms of final and efficient causes are particular harmonies.[8] Baumgarten calls these, respectively, the kingdom of wisdom (a final nexus) and the kingdom of power (an effective nexus) (§358: BM 167).[9] Again following Leibniz, the German professor conceived a further harmony between the kingdom of nature in which composites of bare monads comprise moving and resisting material bodies (§§430–433: BM 181–182) and the kingdom of grace that is populated by rational, moral immaterial spirits (§403: BM 176; §434: BM 182).

From the universal nexus, the ordered, rational interconnectedness of all substances in the world, Baumgarten inferred that from any one substance, it is, at least in theory, possible to know all the rest of the universe and concludes therefore that each monad is, as each is for Leibniz, a microcosm mirroring the macrocosm:

> from any given monad of every composite world, and hence of this composite world, one can know the parts of the world to which this monad belongs (§14), i.e. every *monad* of every composite world, and hence of this composite *world*, is a power (§199) for representing *its own universe* (*they are active mirrors* of their universe (§210), *indivisible* (§244), *microcosms, abbreviated worlds, and concentrations of their own worlds*, or they have a power, they are endowed with a power for representing their own universe).
>
> (*Metaphysics* §400: BM 175)

Using Leibniz's classification of ideas as obscure, confused, clear, distinct, adequate, symbolic or intuitive (*Meditations on Knowledge, Truth, and Ideas*: GP IV 422–426; AG 23–27),[10] Baumgarten maintained that, through their confused sensations, monads represent 'the present state of the world' from the standpoint of their own bodies (*Metaphysics* §534: BM 205). However, when he considered the 'bare' or 'slumbering' monads that 'represent this world either only obscurely, or at least partially clearly' (§401: BM 175), Baumgarten appears to have parted company with his predecessor. For Leibniz, each monad – or each corporeal substance that is a dominant monad together with its aggregate organic body – is an indivisible *metaphysical* point. *Physical* points, on the other hand, are microscopically minute, divisible, extended organic bodies. No matter how small they appear relative to us, physical points are always aggregates of monads or corporeal substances whose organic bodies are in turn further divided (*New System*: GP IV 483; AG 142).[11] Leibniz's physical points are composite pieces of secondary matter, possessing the derivative active and passive forces manifested in the physical points' motions and resistances.[12] On his side, Baumgarten endorsed Leibniz's view that the bare monads are immaterial, indivisible powers or forces; that they are the ultimate elements of material extended bodies; and that they come together to form 'derivative corpuscles' that have other corpuscles as their parts (*Metaphysics* §426: BM 180). He agreed too that material bodies are aggregates-of-substances (i.e. pieces of secondary matter) with inertia and motive power (§416: BM 176; §418: BM 177). However, because the bare or slumbering monads are 'those from the aggregate of which an extended being arises', Baumgarten is willing to allow that they be conceived as physical points (*Metaphysics* §399: BM 175). In this respect, they are akin to what he later describes as 'primitive corpuscles', that is, simple corpuscles that have no parts, but which combine as elements to form bodies, or 'derivative corpuscles' (*Metaphysics* §§421, 426: BM 179, 180). Thus, instead of distinguishing metaphysical points (monads) from physical points (minute bodies), as Leibniz had done, Baumgarten's 'bare' monads themselves become the physical points from which bodies arise (*Metaphysics* §399: BM 175).

Since every Leibnizian monad has an organic body over which it is dominant and with which it comprises a living corporeal substance, and given that each organic body, on the death of the corporeal substance itself, contracts into an invisible physical point, it is understandable that Baumgarten conceived the always-embodied bare monads (rather than these monads' organic bodies) as actual physical points. However, although the difference is small, it is significant, for in conceiving gross bodies as compounds of bare (physical-point) monads, the fact that the bare monads are soul-like entelechies that have perceptions and appetitions is then allowed simply to recede into the background. Baumgarten's stress on the physicality of the bare monads thereby set the scene for Kant's pre-critical dynamical physical monadology in which monads are nothing more than attracting and repelling physical forces and the notion of bodies as aggregates of living ensouled corporeal substances plays no part.

Kant and post-Kantian German philosophy

Immanuel Kant (1724–1804) had been instructed in the Leibnizian philosophy while a student at the University of Königsberg. It has been suggested that his teachers at the university may have included the Wolffians, Conrad Theophil Marquardt and Carl Heinrich Rappolt (Kuehn 2001, 16). Outside the university, Kant may also have encountered Christian Gabriel Fischer (1686–1751) who had returned to Königsberg in the mid-1730s, following his expulsion from the University and the city in 1725 on account of his admiration for Wolff's philosophy. However, when Kant himself assumed the role of teacher at his alma mater in 1755, it was to Baumgarten rather than Wolff that he directed his students, for it was Baumgarten's Leibniz-inspired *Metaphysics* that Kant used as the basis for his lectures throughout his teaching career (Cassirer 2009 [1951]: 338; BM 3, 22, 54). The copious written notes he made in his own copy of Baumgarten's text evidence Kant's engagement with the text and its influence on the development of his own ideas (BM 25).

Kant's first published book was his *Thoughts on the True Estimation of living forces and assessment of the demonstrations that Leibniz and other scholars of mechanics have made use of in this controversial subject, ...*(1746–49).[13] As the title indicates, Kant here addresses the question of the living force (*vis viva*) in bodies that Leibniz had postulated as an alternative to the inert passive matter favoured by the Cartesians. Leibniz had published a paper in the *Acta Eruditorum* of March 1686 – *A Brief Demonstration of a Notable Error of Descartes and Others concerning a Natural Law* – in which he had argued that the quantity of motion of a body does not serve as a guide in estimating the strength of its force, since, contra Descartes, the quantity of motion in nature is not conserved, but the quantity of motive force is conserved. Leibniz's mature thoughts on dynamics appeared in 1695 when he published the first part of his *Specimen Dynamicum* in the April issue of the same journal. There, Leibniz explained and defended his physics of living and dead forces. Kant's *True Estimation* is his contribution to the (in)famous '*vis viva* controversy' over the presence of living forces in bodies that had ensued since the publication of Leibniz's *Brief Demonstration* in 1686.[14]

The living forces described in *Specimen Dynamicum* are not the primitive active forces that are the souls or substantial forms in corporeal substances (*Specimen Dynamicum*: GM VI 236; AG 119). Rather, they are the derivative active forces in composite extended bodies and whose presence is known by their effects, the motions of a body or the motion of its parts (GM VI 236–237; AG 119–120). That these physical derivative forces are ultimately modifications of the same monadic primitive forces that give rise to monads' perceptions and appetitions is a consideration in metaphysics, but not in dynamics. Souls and forms – primitive active forces – cannot intelligibly be invoked to explain 'the individual and specific causes of sensible things' (GM VI 236; AG 119). The science of dynamics leaves aside metaphysical speculations about the existence of souls and substantial forms in bodies and restricts itself to mathematical and empirical study of the *derivative* active and passive forces in bodies, that is, to bodies considered only as objects that move and resist. In this context, living forces are so-called, not because they are the forces of living biological creatures, but rather because they are the forces present when bodies are actually in motion. In this respect, they are contrasted with dead force, in which there is only a 'solicitation to motion [*motus*]' but not yet any actual motion in the body (GM VI 238; AG 121).

Kant's *True Estimation* considers the mathematical and empirical evidence that might support the theory of living forces in bodies, along the way acknowledging that 'in order to determine the true estimation of force in nature, we must connect the laws of metaphysics with the rules of mathematics; doing so will fill in the gap and better meet the designs of God's wisdom' (*True Estimation* §98: AK I 107; Watkins 2012, 94). In this work, however, while not denying living forces (§50: AK I 59; Watkins 2012, 56), Kant expressly refrains from entering into any metaphysical speculations about the entelechies that might provide their philosophical grounding (§16: AK I 28; Watkins 2012, 31; §62: AK I 70; 65; §90: AK I 98; 87).

Kant picks up the issue of corporeal forces again in his *Physical Monadology*, published in 1756. The 'monads' that Kant introduces here do bear some resemblance to Leibniz's monads insofar as they are simple, indivisible substances, compounds of which comprise bodies (*Physical Monadology*: AK I 477; Walford 1992, 53), but his description of them is also very much in keeping with the bare monads or physical points of Baumgarten, for Kant's physical monads are centres of attractive and repulsive physical forces. For Kant, these forces constitute the monad's 'sphere of activity' by which it fills a space that 'prevents the monads immediately present to it on each side from drawing closer to each other' (AK I 480–481; Walford 1992, 57). The repulsive forces ensure that a body compounded from simple monadic physical elements is impenetrable, while the monads' attractive forces ensure that the impenetrable body also has volume. Together, volume and impenetrability define the extent or 'limit' of the body's extension (AK I 482–484; Walford 1992, 59–62).

As we saw in chapter 2, the activity or modification of the Leibnizian monad gives rise on the one hand to its internal modifications, namely, its perceptions and appetitions, and on the other hand to its external modifications, its derivative forces. The modification of the monad's primitive forces as derivative forces requires the creation of the subordinate monads that comprise its extended organic body.[15] It is through thus possessing an organic body that the monad can be said to have a position in space, even though it itself does not 'occupy' or 'fill' space. In contrast, the activity of the Kantian monad generates the sphere of activity by which the monad itself actually fills space. Moreover, in Kant's physical monadology, the monad's perceptions and appetitions, downplayed but not eliminated by Baumgarten, have now been completely eradicated. Kant had recognized souls that have representative natures in the *True Estimation*. There, souls are said to be capable of being united and of interacting with their bodies through their representations (*True Estimation* §6: AK I 21; Watkins 2012, 25), but the question of perceiving entelechies in bodies is left unresolved. In the *Physical Monadology*, however, even this question is ignored and the monads that comprise bodies are not assigned any perceptual or representative capacities at all. Ultimately, Kant would come to recognize the need to reintroduce some kind of appetition and final teleology into bodies in order to account for the ordered complexity of living organic bodies (chapter 4), although he still drew back from fully committing to the presence of perceiving, appetitive souls in living bodies.

Post-Kantian German philosophy saw the appointment in 1809 of Johann Friedrich Herbart (1776–1841) to a professorship at the University of Königsberg, holding the same Chair as had Kant. Herbart's monadological metaphysics maintains that the existence of a plurality of simple, unextended, independent or self-sufficient unities or 'reals'. Intentionally fashioned on Leibniz's monads, Herbart conceived the reals as active living forces that underlie the phenomenal qualities of bodies. All the same, Herbart eschewed Leibniz's teleology, preferring to understand the reals in terms of their mechanical interactions.[16] However, in the philosophies of Johann Gottfried Herder (1744–1803), Friedrich Wilhelm Joseph Schelling (1775–1854), and Rudolph Hermann Lotze (1817–1881), the perceptual and appetitive aspects of the constituents of the natural world resurfaced, for each of these thinkers in their own ways developed ideas of individuals as active indivisible unities or perceiving, appetitive organisms that bear close affinity to Leibniz's indivisible corporeal substances. Each focused on the whole living organism, conceiving it in terms of active dynamic force. Herder, for instance, acknowledged with Leibniz that the soul is always attached to an organic body through which it comes to know the external world.[17] Schelling, meanwhile, was attracted by the idea of a 'natural monad' as an 'original actant' (Schelling 2004 [1799], 21) and despite his tendencies towards Spinozistic absolute monism,[18] he championed a multiplicity of such monads as dynamic atoms or immaterial vital forces,[19] that, while not parts of matter, must nevertheless be posited as the ideal original grounds of the seeds or 'qualities' in Nature and which, abstracted from their 'products' (namely the organic bodies themselves) do not exist.

> The philosophy of nature assumes 1) with *atomism* that there is an original multiplicity of individual principles in Nature – it brings multiplicity and individuality into Nature with it. – Each quality in Nature is a fixed point for it, a seed around which Nature can begin to form itself.
> (Schelling 2004 [1799], 21)[20]

Later, Lotze would conceive a world comprising centres of force that are almost indistinguishable from the Leibnizian monad as described earlier in this chapter, for, as Beiser explains, on Lotze's reading, 'Leibniz makes Spinoza's two attributes, thought and extension, into properties of each monad rather than a single infinite substance' (Beiser 2013, 146). Just as the monad's primitive active and passive forces are modified simultaneously as its appetitions and perceptions and as the derivative forces of its organic body,[21] such that the monad not only represents the world but also acts through its body to effect its goals and aspirations, so too Lotze's individuals are psychophysical agents, spirits that 'have the power to act for the sake of ideas and to realize them in the world' (Beiser 2013, 269). However, Lotze's Leibnizian pluralism is ultimately resolved into a Spinozistic monism, a position to which Schelling had also been drawn.

Pierfrancesco Basile has argued that in Lotze, monism is the inevitable outcome of his rejection of Leibniz's doctrine of pre-established harmony. He explains that, for Lotze, pre-established harmony wholly undermined freedom of the will: monads' perceptions can only unfold in accordance with God's plan, from which the monads could not deviate.[22] Besides, monads merely duplicate the world that already exists in God's mind[23] and such duplication is wholly unnecessary: 'If creation involves that God has a complete intuition of the world, nothing justifies the creation of a duplicate: the world may well exist as an internal articulation of God's Mind, all existing things could just be His "ideas" or "thoughts"' (Basile 2006, 32).[24]

Herder, Herbart[25] and Schelling[26] had also rejected Leibniz's doctrine of pre-established harmony.[27] According to Herder, '[t]he philosophy-of-pre-printed-forms [*Formular-Philosophie*] which unwinds everything from out of itself, from out of the monad's inner force of representation' offers no explanation of how what is unwound came to be in the soul in the first place (*On the Cognition and Sensation of the Human Soul*: Herder 2002, 208). On the one hand, the soul's representation of the external world is possible only if the soul and body actually interact with each other and on the other hand, the soul can effect changes in its body only if there is causal interaction between them (*On the Cognition and Sensation of the Human Soul*: Herder 2002, 195). Through his analysis of Herder's 1769 *Über Leibnitzens Grundsätze*, DeSouza has argued that it was Herder's perceived need to account for causal interaction between the soul and its body that prompted him to postulate the materiality of the monad. Herder simply transforms immaterial monadic force into material force and, thus, by conceiving souls and bodies as having essentially the same nature, makes the notion of their interaction more credible (DeSouza 2012, 788). In effect, Herder adds the soul or life force of the organism to the constant flux of its moving and resisting body, allowing information to pass from body to soul and from soul to body and removing the need to postulate a harmony pre-established between them.

French spiritualism and British idealism

Aside from Madame du Châtelet whose 1740 *Institutions de physique* included an enthusiastic exposition of Leibniz's thought (Barber 1985 [1955], 135–140), Leibniz had few supporters in France in the eighteenth century. His thought suffered from near total obscurity tempered by unfair and unfavourable criticism based on superficial knowledge of his views.[28] By the nineteenth century, the climate had changed dramatically for the better. Jeremy Dunham attributes the revival in France of Leibniz's monadological pluralism to François-Pierre-Gonthier Maine de Biran (1766–1824), who had published an *Exposition of the Philosophical Doctrine of Leibniz* in 1819.[29] This treatise, Dunham argues, 'was in part responsible for a significant change of direction in French philosophy and its influence can be recognised in a lineage that passes through Félix Ravaisson, Pierre Leroux, Émile Boutroux, Henri Bergson, to Gilles Deleuze' through its emphasis on, and development of, Leibniz's own insistence on the fundamental philosophical importance of a pluralism of self-acting individuals and correlative notions of activity, force and of embodied and introspective experience (Dunham forthcoming, *passim*).

Maine de Biran himself had no sympathy for Leibniz's signature doctrine of pre-established harmony, but this did not deter later French spiritualists, such as Félix Ravaisson-Mollien (1813–1900) and Charles Bernard Renouvier (1815–1903) from regarding the doctrine more favourably. Intimations of pre-established harmony would seem to lie behind Ravaisson's vision of the world as operating simultaneously and harmoniously both by logical, absolute or geometrical necessity and by moral necessity or *convenance* (fitness). In his 1867 report on the state of philosophy in nineteenth-century France, Ravaisson approvingly noted that for Leibniz, 'there is geometry even in the moral and the moral even in geometry' (Ravaisson 1885 [1867], 268). The remark follows on immediately from Ravaisson's description of the sage whose choices are at once both infallible or certain and yet also truly free, being grounded in love of the best and most beautiful, a description that accords perfectly with Leibniz's own account (chapter 6):

> The sage, in choosing the best, chooses it infallibly, [and] at the same time with the most free will. It is perhaps that the good, or the beautiful, is in reality nothing other than love, which is the will in all its purity, and that to will the true good, is to will it [in] itself.
>
> (Ravaisson 1885 [1867], 268)

Just as for Leibniz, so too for Ravaisson teleology reigns supreme. All things tend towards perfection, good, and beauty (Ravaisson 1885 [1867], 270). Even those beings that inhabit the darkest regions of the physical world have 'a kind of obscure idea of the good and of beauty' (Ravaisson 1885 [1867], 271) such that the movements of their bodies might be explained in terms of an inner spontaneity that propels them towards what they perceive as good. Thus, 'as Leibniz said, what one calls physical necessity is a moral necessity that excludes nothing' (Ravaisson 1885 [1867], 271). Accordingly, Ravaisson's vision of the world's progress towards perfection is thoroughly Leibnizian (chapter 8):

> nature offers everywhere a constant progress from the simple to the complex, from imperfection to perfection, from weak and obscure life to life more and more energetic, more and more intelligible, and intelligent [in its parts] taken together.
>
> (Ravaisson 1885 [1867], 269)

Renouvier's Leibnizianism was even more pronounced. The opening sentence of his *New Monadology* powerfully recalls the opening of Leibniz's *Monadology*: 'The monad', according to Renouvier, 'is the simple substance' which has to be assumed since composite substances exist (§1: Renouvier and Prat 1899, 1). The sections that follow continue in the same manner to evince a remarkably close affinity between the monads of Renouvier and those of Leibniz. Monads are 'logically complex', having qualities (§1), but unextended and indivisible. They are, as Leibniz held, the 'true atoms of nature' (§3), the elements of composite substances, whose essences consist of active forces that spontaneously give rise to their modifications (§8), that is, to their perceptions (§7) and appetitions (§9),[30] which vary over time as they reflect the changes occurring in others (§7) from their own point of view (§16). Interestingly, Renouvier was mindful of the relational nature of the monads' identities. He appreciated that monads' perceptions are internal relational qualities. The monad, he observed, is 'a subject of subjective, internal relations' (§4) that is also related externally to other monads (§5). Renouvier also recognized the role these relational qualities play in grounding the appearances of space and time (§§12–13).[31] Furthermore, while Renouvier was cognizant that were we to consider only the representational content of monads' perceptions, we would be unable to distinguish one from another (§6), he agreed with Leibniz that in fact each monad is a unique individual, for no two monads perceive this content with the same degree of distinctness (§12).

Although Renouvier also approved the doctrine of pre-established harmony (§§21–23), describing it as the 'dynamical order of mutual dependence of representations of the monads as active and passive' (§25: Renouvier and Prat 1899, 29), it is also at this point that the novelty of Renouvier's *New Monadology* becomes explicit, for the French spiritualist merely postulated and did not confirm the existence of an external world of bodies – as aggregates of monads – whose order runs parallel to the internal ordering of our own perceptions. Pre-established harmony must rather be assumed as a practical tool to allow us to generalize from our own internal order to a more objective or universal order of the world at large. It is therefore quite erroneous to regard pre-established harmony as the 'negation of causality', for in fact, 'pre-established harmony is the law of natural consequence, the law of nature which regulates psychical and physical causation' (§22: Renouvier and Prat 1899, 21–22). As he noted in his *Principles of Nature*, professors of philosophy had in the past regarded Leibniz's pre-established harmony as bizarre, but what is really absurd is 'to imagine that it must be possible to explain how one thing is the cause of another thing' (Renouvier 1912 [1864], 15). Pre-established harmony does not explain causality. On the contrary, causal communication just is the 'harmony of phenomena [i.e. monadic representations] in time' (Renouvier 1912 [1864], 15–16).[32]

Philosophy in Britain did not see a renaissance of genuine interest in Leibnizian pluralist metaphysics until the late nineteenth and early twentieth centuries, although it is clear that some philosophers, such as Hume, had been reading Leibniz much earlier.[33] As in France,[34] so too in Britain, Leibniz's rationalism had fared badly in competition with Lockean empiricism and its tendency to materialism, but the situation was no doubt exacerbated in the British context by the insular nationalist reaction against Leibniz in the aftermath of the priority dispute with Newton. English sensitivities had abated by the end of the nineteenth century and access to Leibniz's writings was greatly enhanced by the publication of Carl Immanuel Gerhardt's seven-volume edition between 1875 and 1890. Bertrand Russell's *A Critical Exposition of the Philosophy of Leibniz*, based on a series of lectures on Leibniz he delivered at Trinity College, Cambridge, was published in 1900.[35] The volume included a lengthy appendix of key passages on which Russell had based his interpretation, nearly all of which Russell had translated directly from Gerhardt's edition.

Both Russell (1872–1970) and, in France, Louis Couturat (1868–1914)[36] emphasized the logical aspects of Leibniz's philosophy. However, Leibniz's monadology would soon be catapulted back into vogue with the publication in 1911 of James Ward's 1907–10 Gifford Lectures, *The Realm of Ends, or Pluralism and Theism*. In these lectures, Ward (1843–1925) defended a pluralism of interacting monads, building on his references to Leibniz in his earlier 1896–98 Gifford Lectures, published in 1899 as *Naturalism and Agnosticism*, in which he had argued against materialist philosophies and defended a panpsychist experiential ontology. Arguably, Ward's interest in the Leibnizian monadology was sparked when he attended lectures of Lotze and Trendelenberg in Germany and further encouraged by his engagement with the work of Renouvier.[37] Certainly Ward and Renouvier agreed that the monad is a simple experiencing substance that serves as the subject of qualities, and both espoused pluralism and correspondingly rejected monism. However, Lotze's reworking of Leibniz's monads as interacting substances may have proved the greater influence on Ward insofar as Ward, like Lotze, rejects Leibniz's doctrine of pre-established harmony and advocates in its place a pluralism of 'monads with windows' (Basile 2006).

Ward had taught Russell at Cambridge, but it was the elder coauthor of Russell's *Principia Mathematica*, Alfred North Whitehead (1861–1947) who took forward Ward's notion of the interacting monads. Under the influence of both Ward and American pragmatist, William James (1842–1910), Whitehead developed his speculative system of philosophy in which the basic units are experiencing centres of force or 'actual occasions' in the course of his 1927/28 Gifford Lectures, with the published version appearing as the work with which his name is all but synonymous, *Process and Reality: An Essay in Cosmology*, in 1929. There is much to be said for thinking of the Whiteheadian actual occasion as a type of Leibnizian monad. After all, the occasions are active experiences that combine into aggregate bodies or nexuses. The Whiteheadian organism is a nexus or aggregate of actual occasions, one of which is dominant, in much the same way as the Leibnizian organic bodies are aggregates of monads, one of which is dominant over the subordinate monads, uniting them into a single living corporeal substance. So too, actual occasions themselves are described as having both psychical and physical aspects, i.e. mental (or conceptual) and physical poles (Whitehead 1960 [1929], 366). However, Basile warns against understanding the physical pole as an actual physical body. It is, he argues, best understood as the physical aspect of what is still essentially an experience (Basile 2009, 10–11). This is perhaps not too far distant from the way in which for Leibniz monads' experiences or perceptions admit of varying degrees of confusion and distinctness, with the former associated with sensory perception, the higher degrees of the latter with intellectual cognizance.

Understanding the convergence of Leibniz and Whitehead in this manner suggests strongly that the actual occasion is more akin to a monadic perceptual state than it is to the monad itself. The same conclusion can be drawn from the key point on which Whitehead diverged dramatically from the Leibnizian world-view. Essentially, Whitehead took issue with the substantiality of Leibniz's monads, as well as the associated subject–predicate logic that maintains that substances are the bearers of qualities, qualities that are attributed as predicates to a subject that is always a subject and never itself a predicate. Stated simply, Whitehead rejected the notion of a substance that persists through change. For Whitehead, the changing states are all that there is. There is no substance-monad that unfolds its essence as a sequence of different experiences or perceptions. Instead, there are only temporary fleeting experiences. This said, one can view actual occasions as the Whiteheadian counterparts of Leibnizian monads' appetitive–perceptual states. The experiential states that follow on from one another in accordance with the monad's law of the series of its perceptions are, like Whitehead's actual occasions, experiences that both draw on the past and strive towards the future (chapter 8). Actual occasions are fleeting or momentary experiences, constantly in the process of becoming and perishing. Indivisible but complex, they both prehend the past, momentarily synthesizing the past in the present before giving way to the future. In that present moment, they too, hold the universe in a single perspectival view, mirroring the whole universe in a way that forcefully brings to mind the monads of Leibniz, although what they mirror is not, as for Leibniz, the present state of the world, but rather its immediately preceding state (Basile 2009, 13–14).

Having rejected the monads as substances whose essences unfold through time, however, Whitehead was also able to jettison the pre-established harmony, as had Ward before him. If there are no substances in which the future is already in the present waiting to unfold (chapter 8), the future itself opens up as a realm of infinite possibilities. Which possibilities will come to be in the as-yet-undetermined future depends on how each actual occasion prehends its past and on what it fixes its 'subjective aim' towards the future. In this way, novelty and creativity become the defining features of Whitehead's world.[38]

Novelty and creativity are also the hallmarks of the monadology sketched by Whitehead's contemporary Herbert Wildon Carr (1857–1931) in his 1922 *A Theory of Monads* and again in his 1930 *Cogitans Cogitata*. Like Whitehead, Carr was aware that Leibniz's monadology required some adjustments in the light of developments in the physical sciences. Unlike Whitehead, however, Carr did not abandon the notion of the monad as a persisting subject of changing qualities (Carr 1930, 1). Indeed, excluding pre-established harmony, Carr retained the key features of Leibniz's monadology. For Carr, the universe comprises a plurality of monads that are simple, indivisible, perceiving, appetitive active unities whose nature consists of active and passive force. Each monad or entelechy has its own organic body, an aggregate of monads, over which it is dominant and with which it comprises a composite indivisible or unitary organism. Nevertheless, each monad is, in itself, an independent entity, enclosed within the sphere of its perceptions, although its perceptions are always representative of the world outside, mirroring the rest of the universe through the prism of its own organic body, the external expression of its inner activity. Although each monad is a solipsistic individual, unable to represent the universe from anything other than its own point of view, each is also fundamentally relational, incapable of being characterized in the absence of its relations to other monads: 'if I try to conceive my individuality in abstraction from my relatedness to others there is no support for the concept. If I abstract from the relations there is no remainder' (Carr 1930, 32). Following Leibniz closely, Carr regarded these relations as internal to the monad, for there is no actual interaction between the monads:

> My perceptions and actions relate me to other monads not by interaction or interchange but by expressing the mutual influence of my being on them, and their being on me. I am myself within the world of my perceptions and actions, and in that world I am one of many.
>
> (Carr 1930, 29–30)

Thus, Carr's monads are, like those of Leibniz, essentially 'windowless' in the sense that nothing is transferred from one to another. Again in keeping with Leibniz, Carr did not believe that this precludes communication of one monad with another, for he appreciated that a form of communication is made possible by sympathy and harmony among monads. However, Carr restricted communication among monads to the communication between rational minds by means of speech, whereas for Leibniz, expressive communication occurs between all monads (chapter 7).

Carr insisted that only a Leibnizian style monadology could provide a solid metaphysical underpinning to the sciences of his own time: it is the only means by which we can reconcile 'sentient experience and scientific truth' (Carr 1930, 86). Accordingly, he claimed that his monadology does not merely reiterate that of Leibniz, but completes it: the 'new monadology therefore comes not to destroy but to fulfil the old' (Carr 1930, 110), by making it compatible with Einsteinian relativity and quantum theory. The new physics, he argued, requires us to postulate the existence of a plurality of active individuals, each the centre of its own field of force, but with the experiential or perceptual aspects of these living atoms of force brought centre stage. '[B]asic ideas of space, time and matter' had all but disappeared under the principle of relativity (Carr 1930, 101), but a Leibnizian theory of space and time as the order of coexistences and the order of successions could save the day. Under the influence of Henri Bergson's creative evolution, however, Carr maintained that the harmony among monads is not Leibniz's 'artificial' pre-established harmony, but is rather a 'natural' harmony that arises directly from the 'creative impulsion itself' (Carr 1930, 110). The physical world of objects in space and time is a 'construction which perception requires the mind to undertake' (Carr 1930, 70). Thus the activity involved in the unfolding of Leibniz's monads as a series of perceptual states becomes in Carr's monadology the free activity of monads creatively constructing within themselves the world as perceived from their own perspectives. Nothing is pre-established. Just as for Whitehead, so too for Carr the creation of novel future constructions remains open.

The kinds of speculative systems of philosophy discussed in this chapter fell out of fashion within the circles of analytic Anglo-American philosophy of the later decades of the twentieth century,[39] although Leibniz's thought continued to inspire research in specific domains[40] and scholarly research by historians of philosophy since the latter half of last century has greatly increased our understanding of and access to his work. In Germany and France, however, Leibniz's speculations continued to inspire, proving hugely influential in shaping the philosophies of, for instance, Martin Heidegger (1889–1976), Gilles Deleuze (1925–1995) and Michel Serres (1930–).

Leibniz's dynamism as well as his Principle of Sufficient Reason had a profound effect on the philosophy of Heidegger. Hans Ruin (1998) traces the development of Heidegger's reaction to Leibniz's rationalism through first and third of the three sets of lectures on Leibniz that Heidegger delivered at successive points in his career.[41] Crucially, in the first set of lectures, Heidegger raises the question of the reason or ground of the Principle of Sufficient Reason itself (Ruin 1998, 56), which in turn led him to 'an understanding of being which incorporates its own transcendental foundation as the freedom and transcendence of the human being' (Ruin 1998, 58). In other words, for Heidegger, it is part of the very being of Dasein to be 'an existence ... projected toward the world as an object of care and questioning' (Ruin 1998, 60). In chapter 6, we shall find the same to be true of Leibniz's rational monads, but there are other ways in which Dasein resembles the Leibnizian monad. Ruin points perceptively to the inherent relationality and the dissolution of the subject–object distinction of both (Ruin 1998, 59), while Lodge outlines similarities between Heidegger's concept of 'drive' and the representative, forward-looking, unifying, simple and dynamic nature of the Leibnizian monad (Lodge 2015). Leibnizian dynamism and reason also played a major role in the influence of Leibniz on Deleuze. In his doctoral thesis, Alex Tissandier explains how Leibniz's search for consistency and order, epitomized in his Principle of Sufficient Reason, and his fascination with the dynamic force and infinite folding or embeddedness of organic structures perform central, if sometimes opposing, roles in Deleuze's 'creative reading' of Leibniz through which Deleuze effectively constructed a 'new, novel Leibnizian system' (Tissandier 2014, 3).[42] Overall, Tissandier discerns the 'constant presence' of Leibniz in Deleuze's writings (Tissandier 2014, 1). A similar claim has been made in respect of the French philosopher, Michel Serres, of whom Gary Gutting notes that, having devoted his 1968 doctoral thesis to an examination of Leibniz's system,[43] Serres' 'subsequent work can be very fruitfully read as a twentieth-century reformulation of Leibniz's philosophy' (Gutting 2001, 233).

The philosophers discussed in this chapter are indicative, but not exhaustive,[44] of the remarkable range and longevity of Leibniz's influence on Western philosophy. We turn now to consider whether Leibniz's speculations might not yet prove inspirational in the twenty-first century.

Notes

1 Pierre Bayle's critique of Leibniz's system of pre-established harmony in note H to the entry on 'Rorarius' in his *Historical and Critical Dictionary* was published in 1697 (WF 68). Foucher's response appeared in the September 1695 issue of the *Journal des Savants* (WF 41n23).

2 There is debate among commentators as to how far Wolff adopted a thoroughgoing theory of representative monads. For a helpful summary, see Rutherford 2004, 235–236n21.

3 Marius Stan (2013) has argued that Kant's third 'mechanical law of action and reaction is developed, not against the Newtonians, but as a solution to issues he regarded as problematic for Leibniz–Wolffian dynamics. Kant could see no reason why a body at rest should suddenly gain motive force just at the very moment that a body, colliding with the first, loses an equivalent amount of its own motive force.

4 For discussion see Watkins 1998, esp. 140–142 and 1995, 302.

5 And, we should add, by Baumgarten's student and disciple, Georg Friedrich Meier (1718–1777).

6 See Beiser 2009 for an excellent and full account of the influence of Leibniz's philosophy in this regard. Guyer 2014, I 305–418 provides a thorough overview and assessment of eighteenth-century German aesthetics. Also see Wilson 1995b, 467–468.

7 Consequently, Baumgarten also conceived God as outside the created world, an infinite substance distinct from the finite substances that comprise the world (*Metaphysics* §388: BM 172).

8 A modern-day follower of Baumgarten might well consider an ecocommunity as a particular harmony or nexus.

9 Contrary to Leibniz, Baumgarten also admits other kinds of causes, for instance material and formal causes (§345: BM 163).

10 Cassirer (2009 [1951], 342) notes that this paper profoundly influenced all Baumgarten's philosophy.

11 Eric Watkins also draws attention to this passage in discussion of the relation of Baumgarten to Leibniz, while also noting that Wolff was agnostic on the question of the monad's materiality (Watkins 2006, 295). With regard to Leibniz, the *New System* passage is ambiguous. Although it is commonly read as asserting that monadic forms or souls are metaphysical points, Leibniz only states here that metaphysical points are 'constituted by forms or souls', not that they *are* forms or souls. Hence, the passage can equally well be read as asserting that the true '*atoms of substance*', the metaphysical points, are actually corporeal substances, with the constitutive role of the forms or souls understood as the soul's act of unifying itself with its organic body so as to comprise the indivisible corporeal substance. Leibniz also claims here that 'when corporeal substances are contracted, all their organs together constitute only a *physical point* relative to us'. This in turn suggests that physical points are minuscule organic bodies of corporeal substances, not corporeal substances themselves, with the constitutive role of the organs in the organic body understood as their serving as the parts of the divisible physical points.

12 Ultimately, these derivative forces are founded upon the indivisible and nonextended monads, for derivative forces are modifications of monads' souls (as primitive active forces) and primary matter (as primitive passive force).

13 Translated in Watkins 2012, 1–155.

14 Iltis 1971 remains, in my opinion, the clearest exposition of the *vis viva* controversy in its early stages.

15 Phemister 2005, 194–202.

16 Beiser 2015. Nevertheless, Herbart did, like Leibniz, appreciate the fundamental importance of the aesthetics of perfection and harmony, regarding the pleasure that we derive from the perception of harmony as the ground of our moral approval (Copleston 1963, 253–254).

17 Wilson 2010, 302–303. Among other notable points of influence, Wilson draws attention to the Leibnizian tone of Herder's views on progress in his *Letters on the Improvement of Humanity* (Wilson 2010, 305). Wilson describes Herder as a 'passionate and lifelong admirer' of Leibniz (Wilson 2010, 301), while DeSouza asserts that '[i]f Leibniz's philosophy is taken as a whole, ... few eighteenth-century thinkers were more deeply influenced by it than Johann Gottfried Herder' (DeSouza 2012, 773–774).

18 See for example, Schelling 2004 [1799], 28, 49–50, 53–54. Schelling differs from Spinoza, however insofar as he does not consider absolute monism as incompatible with a pluralist dynamical atomism (Schelling 2004 [1799], 53–54, 208).

19 Schelling 2004 [1799], 61.

20 See also Schelling 2004 [1799], 52. On the multiplicity of individuals in Schelling's philosophy, see Foster 2011, 227. However, Schelling's monadic actants are not substances and do not endure – 'Every actant will ... be *a different actant* at every stage' (Schelling 2004 [1799], 33). In this respect, they would seem to resemble Whitehead's actual occasions more than they do the monads of Leibniz.

21 See above, chapter 2, pp. 41–42.

22 We address this concern in chapter 8.

23 Each Lotzean centre of force is also, like the Leibnizian monad, a mirror of the whole universe (Lotze 1888, I 360). For discussion, see Conger 1922, 96–98.

24 Beiser notes the tension in Lotze between pluralism and monism. Discussing Book IX chapter 3 of Lotze's *Microcosmus*, he claims that 'Lotze wants to correct Spinoza's monism with Leibniz's concept of individuality, which, he thinks, will alone give reality and independence to the finite modes of his single infinite substance. The crucial question, of course, is whether he can really combine these doctrines. Leibniz's concept seems to shred Spinoza's single infinite substance into an infinitude of finite pieces' (Beiser 2013, 270).

25 Beiser 2015, 1063 and Copleston 1963, 252.

26 E.g. Schelling 2004 [1799], 46–47.

27 This was also the stance taken in the physical monadology of Czech philosopher, Bernard Bolzano (1781–1848). See Simons 2015, 1078.

28 Barber 1985 [1955].

29 Dunham also notes that Biran's interest in Leibniz was profoundly influenced by conversation with Madame de Staël and by her remarks in her book,*On Germany*, published in 1810 (Dunham forthcoming). Dunham records his debt to Patrice Vermenen's studies on Biran. See, for instance, Vermenen 1987.

30 See also Renouvier 1912 [1864], 12.

31 This is in line with the interpretation we develop in later chapters. In chapter 5, it will be shown that the qualities that Leibniz's monads possess are internal relational qualities that serve as the ground for monads' relations to external things. In chapters 6 and 8, we examine how space and time are founded upon these relational qualities.

32 Dunham observes: 'one of the most interesting points about Renouvier's use of pre-established harmony is how he uses this in order to present a non-reductive philosophy of nature. For Renouvier, there are psychological laws, there are physiological laws, there are physical laws etc. ... but it is not possible to eliminate these laws by any process of reduction of one set to the other. Each domain of science has its own irreducible laws and would lose its explanatory power without them. These laws work together not because some sets are supervenient on others, but because of pre-established harmony' (pers. commun., 27 January 2015).

33 Hume described the correspondence 'between the course of nature and the succession of our ideas' as a 'kind of pre-established harmony' (*Enquiry* 5.2.44: Hume 1975, 54) and his account of the mechanism of sympathy clearly echoes Leibniz's theory of expression (chapter 7).

34 Barber 1985 [1955].

35 Russell 1992 [1900], iii–vi, xvii.

36 Couturat 1901.

37 Dunham pers. commun., 27 January 2015.

38 See e.g. Whitehead 1960 [1929], 248–249.

39 The one exception is that of Timothy Sprigge (1932–2007). His absolute idealism (Sprigge 1984) draws upon the monist philosophies of Spinoza and Bradley.

40 For instance, in mathematics and physics, see Arthur 2014, 201–202 and Davis 2000.

41 On reason in Heidegger and Leibniz, see also Cristin 1998.

42 For an exceptionally clear account of Deleuze's use of Leibniz's notion of the fold, see Lærke 2015.

43 Serres 1968.

44 Deleuze, for instance, remarked on the influence of Leibniz on Nietzsche's perspectivism (Deleuze 1980). The influence of Leibniz on Nietzsche's philosophy of nature and conception of time was indirect, mediated through the physicist, Ruggero Giuseppe Boscovich (Ansell Pearson 2000).

4   Organic and inorganic nature

Anyone who diligently compares the products of nature, wrested from the womb of the Earth, with the products of the laboratories (thus we call chemists' workshops) will accomplish an important task, in our opinion: for then the striking resemblances between the products of nature and those of art will shine before our eyes.

(*Protogaea* §9: Dutens II-2, 209; Rossi 1984, 61)

In this chapter, we draw out some of the ecological implications of Leibniz's opinions on living bodies and their similarities and differences with nonliving bodies. We begin, however, with the observation that while Kant's transformation of Leibniz's monadology into a physical monadology sufficed to explain the operations of nonliving bodies (chapter 3), his failure to acknowledge monads as perceiving, appetitive or goal-directed beings rendered the explanatory power of the physical monadology inadequate in respect of living bodies. When the mature Kant came to consider living natural organisms in the second half of his *Critique of the Power of Judgment*, he had to admit that mechanism alone is not sufficient to account for the formation of living organized beings and that a philosophy of nature must acknowledge the operation of teleological principles as well.

For Kant, natural living things are organized beings that are 'natural ends'. By 'natural end', Kant means something that is both 'cause and effect of itself' (AK V 370–371; CPJ 243). However, the functional organization of parts in a unified whole that is essential to our concept of an organism with 'self-propagating formative power ... cannot be explained through the capacity for movement alone (that is, mechanism)' (AK V 374; CPJ: 246). A natural organism cannot be explained solely through a reductionist appeal to parts that come together purely mechanistically, as if they merely by chance happen to form a unified organized creature. References to the mechanical operations or motions of the body's parts cannot adequately account for the organism's functional unity. Such appeals to efficient causation are not enough. The concept of final causation – of things as ends in pursuit of their own goals – must serve as a 'regulative concept for the reflecting power of judgment' that guides research into natural organisms and allows us to think of them in terms of the kind of final causality we find operative in ourselves (AK V 375; CPJ: 247). We need to introduce a teleological regulative principle that will allow us to employ maxims such as 'that everything in the world is good for something, that nothing in it is in vain' and that leads us to 'expect nothing in nature and its laws but what is purposive in the whole' (AK V 379; CPJ: 250).

Attempts to reductively and mechanistically explain the whole in terms of its parts are destined to fail when applied to natural organisms, for in their case, the parts themselves can be understood as parts only in relation to the whole that they serve. With respect to both their existence and their form, the parts of creatures in nature, of things that are 'natural ends', 'are possible only through their relation to the whole' (AK V 373; CPJ 244–245). The parts are able to organize themselves to form the whole because the whole organized being has not just 'motive power', but also has a 'formative power' that it 'communicates' to its matter (AK V 374; CPJ 246). The parts then come together so as to constitute the living thing itself. In this way, the creature or natural end is indeed both cause and effect of itself: the parts of its organized body are the effects of the creature's formative power, but they are also the causes of that same formative power, causes of the form or organized structure of its natural body. Thus, the organized being is both the final and the efficient cause of itself. Its organs are organs only when understood in relation to the whole, that is, when they are understood as effects of the whole; but the same organs are also causes of the whole insofar as the organs themselves produce, maintain and repair each other, keeping the whole together as a single organized natural end. Thus, to use Kant's own example, a tree is a natural end or organized being that is both cause and effect of itself. The tree has reproductive capacity: in producing seeds, one tree generates another of the same species. It also has the capacity to grow as an individual by taking in nutrients and transforming the ingested material into parts of its own body. The tree also has a self-preserving capacity: the parts produced by the whole assist in the preservation of the whole (for instance, the tree produces leaves that 'preserve it in turn, for repeated defoliation would kill it, and its growth depends upon their effect on the stem') and it is a self-repairing mechanism, such that, for instance, damage to one part is 'made good by the others' (AK V 371–372; CPJ 243–244).

We can detect the shadow of Leibniz's corporeal substances and their organic bodies behind Kant's conception of an organized being. Leibniz gives a detailed account of the structure of the corporeal substance in a letter to his physicist friend, Burcher de Volder. He begins by stating that:

> If you take a mass to be an aggregate containing many substances, you can nonetheless conceive of one substance that is preeminent in it, if indeed that mass constitutes an organic body animated by its primary entelechy.
>
> (To de Volder, 20 June 1703: LV 264–265)

This animating primary entelechy is the primitive active force or soul of the dominating monad. In Kantian terms, it is the self-propagating formative power. In this capacity, the dominant monad acts as the unifying principle that makes itself and its organic body (the aggregate of subordinate monads with their own organic bodies) into the indivisible unity that is the whole animal or corporeal substance. The animal's dominant monad is the primary entelechy together with its primitive passive force or primary matter. This dominant monad is 'related to the whole mass of the organic body' in which 'innumerable subordinate monads come together'. The subordinate monads themselves are not parts of the organic body, but they are 'immediately required for it, and they come together with the primary [i.e. dominant] monad for the organic corporeal substance, i.e., the animal or plant' (LV 264–265). Leibniz summarizes his position thus:

> I therefore distinguish: (1) the primitive entelechy, i.e., the soul; (2) matter, namely, primary matter, i.e., primitive passive power; (3) the monad completed by these two things; (4) the mass, i.e., the secondary matter, i.e., the organic machine, for which innumerable subordinate monads come together; and (5) the animal, i.e., the corporeal substance, which the monad dominating in the machine makes one.
>
> (LV 264–265)

The subordinate monads come together for the organic body-machine, and the organic body-machine in turn comes together with the dominant monad for the sake of the whole corporeal substance. If there were no aggregated subordinate monads, the dominant monad would have no organic body with which to execute its decisions and desires and through which to perceive changes in the external world.[1] However, the subordinate monads also need the dominating monad. Without it, they would not form an 'organic' (organized) body at all; they would not constitute a living organic machine each of whose parts contributes to the self-preserving activity of the whole corporeal substance. Thus, the Leibnizian corporeal substance is, in proto-Kantian fashion, a self-sustaining and self-organizing being that is both cause and effect of itself: cause insofar as its dominating monad brings about the unification required (directing the subordinate monads so that their bodies move in accordance with the needs of the whole) and effect insofar as the dominating monad finds ultimate completion as an actual corporeal substance through the existence of the subordinate monads (and their respective organic bodies) that together comprise the dominant monad's body.[2] Kant's tree and Leibniz's corporeal substances are, as Smith explains of Leibniz, 'quasi-perpetual-motion' machines (Smith 2011, 72) that have the ability to reproduce themselves, to take in nourishment,[3] to maintain, repair and preserve themselves, and to initiate their own motion. Self-organization also has a crucial part to play: the organs of a living thing are functional parts, organized in such a way as to serve the needs and goals of the whole animal or organism and best understood in terms of final causes (Smith 2011, 70).[4]

Whereas Kant consistently refused to speculate on the metaphysical foundations of natural beings' self-propagating formative powers and was reluctant to posit the presence of a soul or dominant monad in natural organized bodies, Leibniz had no such qualms. It is, after all, the soul that experiences its environment and from within which it prefers to pursue some things and ignore others. If living things are governed by final causes, pursuing whatever is perceived as the best, they must have souls or entelechies, for it is their souls that have the perceptions and appetitions that allow creatures to perceive the objects of their desires and to strive towards them.

Thus, Leibniz's organic bodies are not merely *organized* machines; they are also ensouled *living* machines, mechanical *organisms*, as opposed to mere mechanisms. Organisms are living bodies not just in the sense in which *vis viva* is a living force, but also in the sense of being organisms that are dominated by entelechies or souls that feel, sense or think and have appetites, desires or volitions. Souls experience the world through their unified perceptions of the world from the standpoint of their own bodies. The soul unifies the substances in the aggregate, in such a way that the dominant monad and its organic body form a complete animal or animal-like corporeal substance. Without a dominant soul, the organized body would be a mere inanimate (i.e. *soul-* or *anima-*less) object, an insentient, unfeeling, unthinking or perceiving automaton, not unlike the mechanical figures that used to entertain visitors to the Royal Gardens at Saint-Germain-en-Lay. In contrast, organized bodies that are organic, living bodies are *anima*-ted, possessing dominant perceiving and appetitive souls or entelechies. They are not robotic machines. They are instead ensouled bodies of living creatures. Drawing on the examples of the Electress Sophie's dogs, he writes in a letter that was never sent, that her dogs and other animals are machines,

> each animated by their always subsisting unity, which is called the soul, and which is like a center in which every perception is brought together, or rather without which there would not be any perception in the machine, any more than there is in a clock.
> (To Sophie and Duchess Elizabeth Charlotte of Orléans, 28 October 1696/97 November 1696: A I xiii 87; LS 147)[5]

Despite his insistence that living creatures have souls or entelechies, Leibniz thought that soul-like forces should play no role in scientific, dynamical or mechanical, explanations of any particular natural phenomena, including the activities of the organic bodies of living things.[6] Physical mechanisms are 'sufficient to produce the organic bodies of animals, without any need of other plastic natures' (*Theodicy*, preface: GP VI 40; H 64).[7] All the same, explanations in terms of physical derivative forces have limited scope. Efficient causes operating in accordance with the laws of motion and resistance can explain the everyday collisions between bodies, but they cannot explain *why* bodies operate in the way that they do. A complete explanation must resort to metaphysical and theological principles that lead back ultimately to the divine decision to create the best or most perfect, the most varied and most orderly world possible – a choice that includes the decision to create the world whose natural laws will, given the right initial conditions, ensure that organisms develop in such a way that their motions and resistances at all times correspond exactly to the perceptions and appetites of the souls to which they are attached.

It is in this context that Leibniz invokes the doctrine of preformation:[8]

> As for the motions of the celestial bodies, and even the formation of plants and animals; there is nothing in them that looks like a miracle, except their beginning. The organism of animals is a mechanism which supposes a divine preformation. What follows upon it, is purely natural, and wholly mechanical.
>
> (Fifth Letter to Clarke: GP VII 417–418; LC 93)

The point echoes that already made in the *Theodicy*: '[m]echanism' can only 'produce the organic bodies of animals' so long as there is also 'the *pre-formation* already completely organic in the seeds of the bodies that come into existence, contained in those of the bodies whence they spring, right back to the primary seeds' (*Theodicy*, preface: GP VI 40; H 64).[9]

Each seed is an aggregate organic body, dominated by a soul or entelechy that unifies the whole to form the corporeal substance, that is, the whole animal or animate being. Primary seeds are monads' organic bodies in their initial states, at the moment of the creation. By preformation, all of the future development of the world is already contained in these initial seeds and in the entelechies that unify them. All entelechies and souls, including those destined one day to emerge as human souls, 'have been in the seed, and in the progenitors as far back as Adam, and have consequently existed since the beginning of things, always in a kind of organic body' (*Theodicy* §91; GP VI 152; H 172). The seeds have not always had the shape and size they presently possess nor the shapes and sizes that they will later assume. The organic body that an entelechy had at the moment of creation is not the same organic body that it has now. Nevertheless, it was always the case that this particular seed, say, would become, for instance, the body of this particular sheep at a certain point in its development and that another would become the body of this particular dog, and another the body of this particular fly. So too, it was already certain at the moment of creation that particular human organic bodies would develop from particular initial seeds. The seeds of Adam's descendants have not yet taken on human form, but they have been 'preformed and predisposed to assume one day the human shape' (*Theodicy* §397: GP VI 352; H 361). Many transformations of the animal that will one day be human and many changes to its preformed body must occur before it becomes an embryo, foetus and neonate. According to Leibniz, such transformations abound throughout nature. Most are hidden from view, but some we can observe, such as the dramatic transformation of the caterpillar as it pupates and finally emerges as a butterfly.[10]

Once preformed, seeds develop mechanically, changing their relations to others through their own spontaneously generated movement and resistance. As biological organisms, they grow and reproduce, they take in nutrients, expel poisons, self-repair and eventually decay. Organic bodies are thus in constant flux. As the animals themselves grow or change shape and form,

> they acquire and leave behind only parts. In nutrition this happens a little at a time and by small insensible particles, though continually, but it happens suddenly, visibly, but rarely, in conception or in death, which causes animals to acquire or lose a great deal all at once.
>
> (*Principles of Nature and of Grace* §6: GP VI 601; AG 209)[11]

At first too small to be seen, some will over time become the bodies of visible animals:

> There are small animals in the seeds of large ones, which, through conception, assume new vestments that they appropriate for themselves, which give them the means to nourish themselves and grow in order to pass to a larger stage [*théatre*] and to bring about the propagation of the large animal.
>
> (*Principles of Nature and of Grace* §6: GP VI 601; AG 209)[12]

Thus are the seeds of things that have yet to be augmented and developed into visible animals already present in matter, embedded in the physical world, in our own bodies, in the bodies of other creatures, and in the bodies of creatures strewn throughout rocks and stones and other inanimate lumps of matter. Each individual bacterium in the soil and in our gut, each particle in the air, each atom in the water molecule will be transformed in time. Some will be destined for transformation into larger plants and animals, such as trees, whales, eagles, sheep, cows, wolves, humans and so forth, for as Locke so graphically remarked, 'that, which was Grass to Day, is to Morrow the Flesh of a Sheep; and within few days after, becomes part of a Man' (*Essay* 3.3.19).

The same point expressed in Leibnizian terms may be illustrated with the aid of a mundane example of a woman, Mary, who eats a plate of salad. Under Leibniz's metaphysical system, it would seem to be at least in principle possible that the dominant entelechy of a molecule in a piece of lettuce in Mary's salad could become in time the dominant entelechy or soul of Mary's as yet unborn son.[13] The constant intermingling of bodies in flux – both among the internal parts of a body and between the body itself and those external to it – would suffice to bring about the mechanical changes required to effect such a change in the entelechy's organic body from lettuce molecule to human form.[14] Living things are independent substances, but the boundaries between their organic bodies are fuzzy and constantly changing as parts (the smaller organic bodies of the component living creatures) are taken in and expelled. The very air we breathe is a constant reminder of the blurring of the inner–outer distinction. Every time we eat, what was outer becomes inner. And because every living body contains traces of its past history, holding in the present the combination of all its contributory causes (Phemister 2015), when we consume the other, we take into our bodies not only the food itself, but also the record and the effects of its production. In his novel, *Eating Animals*, Jonathan Safran Foer explores the ethical dimensions of this in relation to the consumption of meat (Foer 2010),[15] but Foer's observations have wider application under a Leibnizian metaphysics, for on this view, fish and plants too are, like all individual substances, living creatures. A potato is the seed of a potato plant and like the potato plant itself, the potato is an aggregate of other smaller living corporeal substances. When we eat a potato, we eat the 'seed' of a potato plant that is then transformed to become a part of our own body. In this extended sense, everything we eat is a case of 'eating animals'.[16]

Let us assume, then, that it is at least in theory possible that the dominant entelechy of a lettuce molecule that is taken into a woman's body when she eats a salad might, together with its now suitably remodelled organic body, come to be located in her reproductive organs and eventually augmented and transformed as embryo and foetus, and expelled in turn as a human baby. However, it is not only the pregnant woman who holds another living being in her body; all living things hold an infinity of living beings in their bodies.[17] Every organic body is an aggregate of corporeal substances, of monads with organic bodies.[18] Each of these organic bodies is in turn an aggregate of monads, again with their organic bodies, and so on in a never-ending infinite nested series of organic bodies composed of organic bodies that in turn are composed of smaller organic bodies.[19] Each organic body is a natural machine that has 'a truly infinite number of organic parts (*organes*)' (*New System*: GP IV 482 post-publication revision; WF 16).

> [I]f matter is arranged by divine wisdom, it must be essentially organized throughout and ... there must thus be machines in the parts of the natural machine into infinity, so many enveloping structures and so many organic bodies enveloped, one within the other.
>
> (*Considerations on Vital Principles and Plastic Natures*:
> GP VI 544; L 589)

Leibniz believed the findings of investigative research in microscopy provided additional empirical support for his view.[20] Of course, the discovery of tiny creatures in matter, invisible to the naked eye, did not prove that their living bodies have infinitely many parts, but it did suggest that the parts uncovered to date are living, self-moving compound bodies and offered the hope that technological refinements of the instrumentation would lead to the discovery of indefinitely many more organized living bodies enveloped within relatively larger bodies.

Leibniz saw in this a way to characterize both the similarities and the differences between the natural and the artificial in a manner that testifies to the wonder and magnificence of the living things created by God, without denigrating human attempts to approach some semblance of their beauty and value in the design and construction of inanimate objects:

[E]ach organized body of a living being is a kind of divine machine or natural automaton, which infinitely surpasses all artificial automata. For a machine constructed by man's art is not a machine in each of its parts. For example, the tooth of a brass wheel has parts or fragments which, for us, are no longer artificial things, and no longer have any marks to indicate the machine for whose use the wheel was intended. But natural machines, that is, living bodies, are still machines in their least parts, to infinity. That is the difference between nature and art, that is, between divine art and our art.

(Monadology §64: GP VI 618; AG 221)

The brass-toothed wheel of the clock is one of a finite number of parts of this artificial machine of human design and construction. These parts are not living bodies. They are not organic bodies dominated by a unifying entelechy. The clock exists only for as long as the finite parts exist together organized as a functionally working timepiece. Separate the parts and the clock is destroyed. The clock has no dominant entelechy to ensure its continuing identity. In this, it is not unlike the rocks and stones, clay, wood, sand and water that builders use to construct houses, or like the brass and steel used in the construction of the clock, or like the copper, zinc and iron that make up these alloys.

Ultimately, however, even these inanimate materials and the artefacts made from them are aggregates of living substances. Hence, Leibniz likens a marble tile to a flock of sheep (to Arnauld, 28 November/8 December 1686: GP II 76; LA 94). Just as there can be no flock of sheep without the individual sheep, so there can be no marble tile unless there are living creatures making up this inorganic aggregate. None of these living creatures is destructible. The infinitely many 'enveloping structures' of organic bodies ensure that it is impossible to 'destroy entirely an animal which already exists' (*Considerations on Vital Principles and Plastic Natures*: GP VI 54; L 589). Although their gross bodies may be dismantled, Leibniz contends, something of the body remains, even if it is too minute to be seen by the naked eye.[21] Even when the sheep dies, its entelechy or dominant monad retains some portion of its organic body. The animal's body, with its infinitely embedded, nested parts, cannot be utterly destroyed.[22] The sheep's body can only be transformed, perhaps into a shape in which it is no longer recognizable as a sheep at all. Although the particular individual substances that comprise the animal's body may come and go, the organization of the substances in the body is always in accordance with the unfolding essence of the animal's dominant entelechy or soul, thus ensuring not only the continued existence of the entelechy but of the animal itself.[23]

Organic bodies' infinitely many enveloping structures ensure too that no finite living creature can ever 'produce any organic body entirely anew and without any preformation' (*Considerations on Vital Principles and Plastic Natures*: GP VI 54; L 589). We cannot create new life. We cannot create the infinitely nested complex bodies that living things possess. We can never create anything entirely from scratch. All we can do is alter and rearrange pre-existing materials, that is, reconfigure already existing aggregates of active, self-moving living beings or corporeal substances. All the same, in these acts of making, reconfiguring and reconstructing, humans – and we might add other living things, such as beavers building their dams and lodges – imitate the divine art:

> [M]inds are also images of the divinity itself, or of the very author of nature, capable of knowing the system of the universe, and of imitating something of it through their own smaller-scale constructions, each mind being like a little divinity in its own sphere.
>
> (*Monadology* §83: GP VI 621; M 31)[24]

Human art 'often imitates nature' (*Theodicy* §147: GP VI 197; H 216).[25] Indeed, it may be said, human art oftentimes extends far beyond mere imitation. For centuries, humans have been altering living things themselves, as for instance in the grafting of trees or the cross-breeding of dogs and other animals and plants. In recent years, advances in synthetic biology have taken these activities to dizzying heights of sophistication. However, even in these cases, the original material is still entirely 'natural' and something of it always remains in the creations we humans make from it. When the synthetic biologist creates new sequences of DNA, these are in essence nothing more than reconfigurations or recombinations of already existing DNA. From a Leibnizian perspective, the synthetic biologist does not create a 'new' organism so much as assist in the transformation of a pre-existing creature. Leibniz reckoned that the soul of an insect that is cut in two will 'remain only in one part ... which will always be as small as is necessary to be sheltered from whoever tears or scatters the body of this insect' (to Arnauld, 30 April 1687: GP II 100; LA 125–126). We might surmise that he would similarly have supposed that the entelechy or dominant monad of a creature already present in one of the strands of DNA of a manipulated DNA sequence will become the dominant entelechy of a newly engineered organism. All that has happened is the unfolding and transformation, with human assistance, of an existing living thing, whose preformed body now assumes a previously unmanifested shape and size. As a persisting living being, the organism retains its individual life force and capacity for spontaneous motion. Once formed, its activities cannot be entirely controlled or predicted. Moreover, as a living being, it cannot be destroyed in the way we might destroy an inanimate building, machine or artwork.[26]

Clearly, synthetic biology has blurred the distinction between divine and human art to a significant degree. Still, one thing is clear: only divine art produces life itself. It is beyond the power of any finite being to create anew or to completely destroy any living thing. All that human (or indeed any other living creaturely) intervention can do is to assist in the development of life in new but already preformed directions. As living beings, there is a case to be made that the products of synthetic biology be granted the same status as all other living things. This brings us to the issue of bioegalitarianism that is taken up in the following section.

Bioegalitarianism

Leibniz's vision of the world is fundamentally biological.[27] As Justin Smith has commented, Leibnizian science prioritized the biological sciences over physics (although these exact terms were not then in use) insofar as the latter is conceived as reducible to the former (Smith 2011, 99). Even though not all aggregate bodies in Leibniz's world are living bodies nor is his world as a whole an organism in the sense that Spinoza envisaged, nevertheless all Leibnizian bodies, whether living or nonliving, are aggregates of living substances and the world itself pulsates as a plurality of perceiving, appetitive, embodied, active, living forces, an astonishingly complex, enfolded, beautiful, infinitely recursive structure of smaller creatures within the bodies of the larger ones. Clearly, Leibniz's nested panpsychism satisfies the 'bio' dimension of ecological 'bio-egalitarianism'.

What of the 'egalitarian' dimension of 'bio-egalitarianism? Will a Leibnizian metaphysics support an ethical norm that requires equal consideration be granted to all living things, a norm that is at least as strong as the 'in-principle' biospherical egalitarianism proposed by Naessian deep ecologists? Certainly, Leibniz's bio-egalitarianism is grounded in a pluralist nominalism that regards each individual member of a species as both unique and valued. Without doubt, every individual being in Leibniz's world is a unique particular being that differs intrinsically from all others. Individual horses are more similar to other horses than they are to sheep and flies, but each horse also has its own distinctive characteristics. Every plant and every flower takes the shape of its own family and type. Two daffodils are more similar to each other than they are to a rose bush. Nonetheless, each daffodil is also different from another daffodil, just as one rose bush differs in precise ways from another. Each is distinctive in its particularity or *haeccitas*. Even the petals on the flowers of these plants have their own peculiar characteristics. One will be a richer shade of yellow than another, or be of smaller size, thinner shape, or it may be minutely different in the ridges on its spine. No two things in the universe have exactly the same physical appearance or internal essence. Leibniz conceived each individual living thing as a lowest or *infima species* that so completely characterizes it as to definitively differentiate it from all others. Effectively, each living thing comprises a species of which it is the only member, for no two substances 'can resemble each other completely and differ only in number [*sole numero*]' (*Discourse on Metaphysics* §9: GP IV 433; AG 41–42).

Every living creature in Leibniz's world is also a thing of value. In the first place, each is, in Kantian terms, a 'natural end' or an 'end in itself'. As vital forces, all strive to preserve their own being. Each has, in a loose sense, an interest in preserving its own life. As an entelechial force, each strives not only to preserve its being, but to perfect it. In Aristotelian terms, each strives to promote its own flourishing, consciously or unconsciously aiming to be as perfect as it is possible for it to be.[28] For the individual, such perfecting takes the form of increasing its power, thereby making its appetitions, desires and volitions more effective and, relatedly, improving the distinctness and adequacy of its perceptions, resulting in its more accurate or truer representation of the world and the other living things that comprise it. The entelechy's organic body is also capable of being perfected in terms of increased (derivative) active force, motion and complexity.[29] Indeed, since the physical states of the organic body exactly mirror the perceptions and appetites of its dominant soul or entelechy, the physical health of the body may be regarded as a reliable indicator of the psychical health of the soul or entelechy. Leibniz proposes such a policy in respect of rational minds and higher angelic spirits, but in chapter 7, we shall extend this to all beings in the context of a theory of empathic communication:

> As for Spirits: since I hold that every created intelligence has an organic body, whose level of perfection corresponds to that of the intelligence or mind which occupies the body by virtue of the pre-established harmony, I hold that a very useful way to get some conception of the perfection of Spirits above ourselves is to think of perfections of bodily organs which surpass our own.
>
> (*New Essays*: A VI vi 307; RB 307)

Every living organic body coexists in harmony with its dominant entelechy, the complex multiplicity and changing states of the body expressed exactly by the corresponding multiplicity of perceptions that comprise the dominant entelechy's successive perceptual states. Reflecting in this way the changes external bodies effect on their own bodies, each entelechy perceives the whole world from its own unique embodied perspective or point of view. Each is thus a mirror of the whole universe: 'each monad is a living mirror or a mirror endowed with internal action, which represents the universe from its own point of view and is as ordered as the universe itself' (*Principles of Nature and of Grace* §3: GP VI 599; AG 207). Leibniz even suggests that in mirroring the world that God created,

> every substance bears in some way the character of God's infinite wisdom and omnipotence and imitates him as much as it is capable. For it expresses, however confusedly, everything that happens in the universe, whether past, present, or future – this has some resemblance to an infinite perception or knowledge.
>
> (*Discourse on Metaphysics* §9: GP IV 434; AG 42)

As mirrors of the universe reflecting the perfection and beauty of God's creation, every living substance is itself a thing of great beauty and is a manifestation of the 'multiplicity-in-unity' that characterizes the Leibnizian rationalist aesthetic. All substances are like 'concentrated worlds' (to de Volder, 20 June 1703: LV 262–263), each one a finite version of the harmoniously ordered yet infinitely varied and complex universe. Of course, to our sensibilities, some creatures seem more beautiful than others. Some may exhibit mental disorders or, like the monkfish, downright physical ugliness. In isolation, they may seem ugly and unbalanced, but all contribute uniquely to the harmony and beauty of the universe as a whole. Taken out of context, we can easily fail to appreciate the monkfish's place in the grand scheme of things, just as we might see only a patch of paint as a 'confused combination of colors, without delight [*delectu*], without art' until we see it as part of a whole painting, at which point we realize that 'what looked like accidental splotches on the canvas were made with consummate skill by the creator of the work' (*On the Ultimate Origination of Things*: GP VII 306; AG 153).

Leibniz even insisted that some ugliness or disorder is indispensable to the perfection of the perfect whole, paradoxically increasing its beauty and perfection. In music, composers often,

> mix dissonances with consonances in order to arouse the listener, and pierce him, as it were, so that, anxious about what is to happen, the listener might feel all the more pleasure when order is soon restored ... On that same principle it is insipid to always eat sweet things; sharp, acidic, and even bitter tastes should be mixed in to stimulate the palate. He who hasn't tasted bitter tastes hasn't earned sweet things, nor, indeed, will he appreciate them.
>
> (*On the Ultimate Origination of Things*: GP VII 306–307; AG 153)

Together, living creatures comprise a rich and varied tapestry to which each creature contributes its unique version of God's creation and bears testimony to God's magnificence: 'the universe is in some way multiplied as many times as there are substances, and the glory of God is likewise multiplied by as many entirely different representations of his work' (*Discourse on Metaphysics* §9: GP IV 434; AG 42).[30] Each is required for the beauty and perfection of the universe and for the glorification of God and his creation. Were any one substance missing, a precious representation of God's work would be missing from the whole. God's glory would be inadequately revered and the resultant gap in the continuum would rupture the perfect order of the universe as the remaining substances' perceptions erroneously mirrored that which was not there.

God himself values all of these living mirrors of the universe[31] and if we truly love God, we will follow his example. The following remark, intended by Leibniz to apply to humans, is easily extended to all living things, including our monkfish:

> one must realize that just as in the best constituted republic, care is taken that each individual gets what is good for him, as much as possible, similarly, the universe would be insufficiently perfect unless it took individuals into account as much as could be done consistently with preserving the harmony of the universe.
>
> (*On the Ultimate Origination of Things*: GP VII 307; AG 154)

In this section, we have extracted from Leibniz's panpsychist pluralism a bio-egalitarian theory that upholds the principle that equal care and consideration

should, in principle, be granted to all living things. The 'in principle' qualification is not optional. Conflicts of interest are common, as when, for instance,
human demand for housing and food leads to the destruction of nonhuman
habitations. If in such cases, decisions are routinely made at the expense of
others, it is not long before the charge of anthropocentrism surfaces, as has
happened in Leibniz's case.

Anthropocentrism

No one would dispute the presence of anthropocentric overtones in Leibniz's
writings. In the *Discourse on Metaphysics*, he instances a king who has to
choose between the life of one of his subjects and the life of a nonhuman
animal, declaring without hesitation that 'all wise persons value a man infinitely
more than any other thing, no matter how precious it is' and that 'we would
praise a king who would prefer to preserve the life of a man rather than the
most precious and rarest of his animals' (*Discourse on Metaphysics* §35: GP IV
461; AG 66–67). In this, the wise king follows divine protocol. Leibniz held that
God is monarch over all rational, self-conscious minds in a kingdom of grace
that excludes other creatures. As 'the greatest and wisest of all minds', God can
'enter into conversation, and even into a society' with other minds. Minds can
'know and love' God and are 'infinitely nearer to him than all other things' that
Leibniz declares 'can only pass for the instruments of minds' (*Discourse on
Metaphysics* §35: GP IV 460–461; AG 66). In the next section, Leibniz explains
further why the happiness of minds is God's 'Principal Purpose'. Their moral
virtues endear them to God and are the basis for a very personal relationship
between the human and the divine. Minds not only express the world in their
perceptions, but they also have (distinct) knowledge of it and conduct themselves by self-consciously and freely acting in accordance with laws of final
causes. In this way, they 'express God rather than the world' and have a nature
'so noble that it brings them as near to divinity as it is possible for simple
creatures', so that 'God draws infinitely more glory from them than from all
other beings' (*Discourse on Metaphysics* §36: GP IV 461–462; AG 67).[32] The
others, on their side, provided the evidence of God's glory in his created works
(*Discourse on Metaphysics* §36), whose 'wonderful beauty' rational beings discover through scientific investigation (*Thoughts on van Helmont's Doctrines*,
first half of October (?) 1696: A I xiii, 51; LS 139).

The sciences, in particular the burgeoning seventeenth-century medical sciences,
had practical uses too. Information obtained through research on nonhuman
animals could be used to help to enhance human health. Justin Smith has drawn
attention to the fact that Leibniz himself, at least in his youth, championed the
practice of vivisection because it might lead to advances in human medicine.[33]
Not without good reason did George Sessions observe that Leibniz valued scientific
knowledge '*primarily* as a technique for material human progress through
mastery and domination of Nature' (Sessions 1977, 507).[34] Picking up on a
remark by Antognazza regarding Leibniz's interests in silkworm cultivation,

Smith also counters Kant's assessment of Leibniz's benevolent attitudes towards nonhuman animals with the hypothesis that a worm for which Leibniz is reported to have expressed solicitous concern could have been one of the farmed silkworms from which he hoped to raise revenue to support the Berlin Society of Sciences.[35] Medical and monetary uses aside, we get a glimpse in the *Theodicy* of Leibniz as a cynical anthropocentrist for whom the nonhuman merely provides food for our thoughts:

> there are in these creatures, devoid of reason, marvels which serve for the exercise of the reason. What would an intelligent creature do if there were no unintelligent things? What would it think of, if there were neither movement, nor matter, nor sense?
>
> (*Theodicy* §124: GP VI 179; H 198)

Despite the anthropocentric rhetoric, however, another side to Leibniz belies such extreme anthropocentrism. Leibniz publicly denounces 'the old and some-what discredited maxim, that all is made solely for man' (*Theodicy* §118: GP VI 169; H 189) and even though virtue – characteristic of minds – is the 'noblest quality', he admits, albeit without examples, that countless other admirable qualities 'attract the inclination of God'. Variety is all-important and if the universe did not embrace nonrational creatures and their particular merits, less good overall would prevail (*Theodicy* §124: GP VI 178–179; H 198). The pre-established harmony of mind and body operates not only on an individual level, but at a global level too. Nature and grace must serve the needs of each other. Teleology governs God's choice of the best possible world and is evident throughout creation as the final causation in entelechies works in harmony with the physical forces and mechanisms of bodies. The net result is that Leibniz's anthropocentrism is balanced with a less obvious but equally robust nature-centrism. Nature must serve grace, but so too, grace is 'in some way adapted to that [the service] of nature, so that nature preserves the utmost order and beauty, to render the combination of the two the most perfect that can be' (*Theodicy* §118: GP VI 168: H 188). The moral kingdom of grace (ruled by God as monarch) and the physical kingdom of nature (established by God as architect) stand in a reciprocal relation of mutual utility: 'nature itself leads to grace, and grace perfects nature by making use of it' (*Principles of Nature and of Grace* §15: GP VI 605; AG 212).

  If grace perfects nature by 'making use of it', nature equally is 'made for us *if we are wise*' (*Theodicy* §194: GP VI 232; H 248, my emphasis). These two claims are closely connected. There are ways to use nature that harm or destroy it and there are other ways that are beneficial both to humans and to nature. The wise will seek to employ the latter. Their distinct perceptions and rational appetites enable them to perceive what is truly good and to align their volitions to God's will. God has willed to create the best possible world, a perfect world of harmonious variety, order and beauty. The wise take pleasure in the beauty and harmony of his creation, but they can also freely and with moral responsibility participate in the perfecting of nature by 'making use' of materials available in the natural world to enhance the perfection of the whole, facilitating and promoting its order and biodiversity and in consequence its beauty. If 'we are wise', nature will indeed be 'made for us', for 'it will serve us if we use it for our service; we shall be happy in it if we wish to be' (*Theodicy* §194: GP VI 232; H 248).[36] If we are not wise, if we fail to allow nature to be perfected in the manner conceived by God, for instance by persisting in the anthropogenic destruction of the natural world and its diversity, then nature will be unable to serve us well[37] and our own happiness will be jeopardized.

On this model, if God prefers the wise and virtuous it is because their self-conscious distinct perceptions of the true good and their rational wills to effect the best possible world make them collaborators in the divine project. In cases of conflict, Leibniz believes that preference should be given to the 'one who loves more generally' on the ground that this person is most likely and best equipped to help others (to Arnauld, early November 1671: GP I 74; L 150). His remark is set in the context of helping people and is not intended as a call to extend benevolence more widely to other biological entities in creation. Nevertheless, the concept does invite such extension (chapter 6), especially if we take seriously the idea of the beauty and perfection of the world as a whole. Wise universal benevolence extended towards every living creature, coupled with knowledge and properly applied technical expertise, benefits the whole of creation in both its physical and psychical aspects. Prioritization of the human is not an open invitation to use the rest of creation for our own selfish ends. Members of the City of God, the kingdom of grace, are expected to assist in the perfecting of nature, to extend care and compassion to all its creatures, encourage biodiversity and help maintain an order that benefits all, or at least as many as possible. Far from being a one-sided anthropocentrism, Leibniz's philosophy has the means to support a fully fledged bioegalitarianism in which the servitudes of nature and of grace are reciprocal and equal and in which a particular human being can be granted special privileges only if or when these privileges are needed for the discharge of his or her moral responsibilities towards others.

Valuing the inanimate

Bioegalitarianism relies on the notion that living things, qua living, are intrinsically valuable. However, unless some value is also attributable to the inanimate, we risk undervaluing ecocommunities, natural and mixed landscapes, as well as built environments and other forms of human art and construction, undermining their crucial role in the formation of the world as the best possible and perfect whole.[38] By its very nature, bioegalitarianism prioritizes the living over the nonliving, favouring those things that are able to experience the world through their feelings, sensations or thoughts and to strive for self-preservation through the satisfaction of their appetites, desires and interests. Thus, bioegalitarianism avoids anthropocentrism only to face a similar charge of biocentrism. Can the Leibnizian philosophy temper this tendency through appreciation also of the inanimate and nonsentient? Might it provide means to justify the attribution of value to inanimate things? Just as we can value each unique animate being, *qua living*, might we not also choose to value each unique inanimate object, *qua existing*? In cases of conflict, the living will, quite rightly, be granted priority over the nonliving, but the nonliving is not utterly valueless. Is *bio*egalitarianism and a degree of biocentrism compatible with a more general *onto*egalitarianism that in principle treats all things, living and nonliving, with care and consideration simply because they *are*, in all their unique particularity?[39]

Of course, Leibniz does not grant inanimate things like rocks and stones and buildings the high ontological status that he awards to living things. Inanimate bodies are phenomena, not substances.[40] However, they are real, well-founded phenomena because they are aggregates of real substances, collections of living corporeal substances.[41] Thus, for Leibniz, even inanimate bodies are not entirely lacking of life, for their constituents are living things. The parts of any inanimate body are the organic bodies of living substances, bodies endowed with their own dominant entelechies. The only difference between the organic body and the inanimate aggregate body is that the latter lacks an entelechy that dominates the aggregate as a whole. It still has entelechies that are dominant over each of its parts, but the aggregate itself lacks the overall unity that an overarching dominant entelechy would provide.[42]

In other regards, inanimate things share many of the features that we value in living things, for instance, their unique particularity, agency, objective beauty and, as we shall see in chapter 5, relational value. We have already noted that each Leibnizian corporeal substance is a unique individual.[43] From this it must follow that every inanimate composite of corporeal substances (rocks, mountains, human artefacts and the like) is also unique and distinctly differentiated from all others. Experience bears this out. When we observe closely the details of particular objects, we find ourselves in the same position as the gentleman in the Herrenhausen Gardens who, no matter how hard he looked, could not find even two identical leaves.[44] There exist no two identical leaves or stones; no two strawberries look or taste exactly the same. Even mass-produced items differ slightly: no two books straight off the press are identical in all respects, for minute changes in print quality would be observable were our senses capable of such fine discriminations. Even though they lack the unity, diachronic continuity and indestructibility of living creatures, even though as real phenomena their existence is in part dependent upon being perceived by God or by some finite creature, and even though they are in constant flux, never exactly the same from one moment to the next, nevertheless, each aggregate body for however long it persists is a unique distinct existent in this created world and to that extent worthy of our attention and respect. On other grounds, such as beauty and utility, we may find reason to value some inanimate things over others, but insofar as each is a unique existent, none is lacking in value altogether.

Inanimate things, environments and materials do lack monads that dominate the whole. Unlike living substances, therefore, they cannot be regarded as 'ends in themselves', following their own aims or goals. However, this does not mean that they altogether lack any degree of agency. All corporeal substances move and resist independently and spontaneously through the modification of their own primitive forces as the physical acting and resisting derivative forces of their organic bodies. Inanimate things are aggregates of corporeal substances and the movement and resistance of the aggregate is simply the combined result of the motions and resistances of its component parts. Thus, as Leibniz explains in *Specimen Dynamicum*, in any collision between bodies, each recedes 'not by force of the other, but by its own force' (GM VI 251; L 448). The idea seems counter-intuitive. We tend to think that inanimate things are only ever moved but do not initiate motion themselves, but because Leibniz conceives bodies as aggregates of living, spontaneously moving beings, he is able effectively also to consider aggregate bodies as things that move, rather than as things that can only be moved.

Jane Bennett has recently mounted a provocative challenge to the traditional view in which she defends the idea that inanimate things, such as a scrap of paper moving in the wind, possess a sort of vibrancy and motion (though not a life) of their own. First proposed in Ja. Bennett (2004) and later developed in *Vibrant Matter: A Political Ecology of Things* (Ja. Bennett 2010), she proposes that 'things' must be brought into the causal nexus and endowed with agency and conceived, following Bruno Latour, not as agents, but as 'actants' capable of influencing events in ways that are beyond human control.[45] She talks of inanimate objects 'calling' to her and refers in this context to Roland Barthes' term 'advenience' to describe the ways in which some objects appear able to attract our attention more than do others (Ja. Bennett 2012). The language is active; objects are regarded as endowed with a degree of agency and power, even though they are still classed as inorganic or inanimate (nonensouled) things.

Bennett is fully aware of the huge significance of her new ontology of things. It involves a complete re-evaluation of the nature of the inanimate that has the potential to revolutionize our relations and interactions with nonliving things. The vibrant nature of the inanimate compels us to recognize its independent agency and our inability to control, direct or predict it. At the very least, this encourages a degree of humility on our part, but if we choose to develop an open and receptive attitude towards all things, we might also experience, as did Bennett, vibrant matter's own demands upon our attention and careful considera-tion. And in freely giving our attention, we may cease regarding inanimate things as mere passive objects present solely for our self-centred gratification and come instead to appreciate, value and respect them for what they are simply in and of themselves.

In a lecture presentation at the New School in New York Bennett focuses on the more passive aspects of inorganic things, exploring the world of inanimate objects from the perspectives of extreme hoarders whose lives have been taken over by the things they hoard. Inanimate objects, she suggests, exude a calming sense of *slowness*, comforting and compensating for human loss by their slow rate of decay: 'thing-power is a power of slowness and its efficacy is in part a function of its exemplary patience, stability, duration'. At the same time, their *porosity* indicates the interchange between inorganic bodies and our own organic bodies, elements of the one becoming elements of the other, all being essentially made of the same basic stuff. And finally, she suggests this last fact enables us to develop a kind of *sympathy* with the inorganic, fostered by a longing in the human body to 'return to the indeterminacy of the inorganic' (Ja. Bennett 2012).

For metaphysical grounding, Bennett turns to Spinoza, but she might equally well have turned to Leibniz. Perhaps even more so than Spinoza's, Leibniz's metaphysics is able at one and the same time to distinguish sentient living and nonsentient nonliving bodies while imbuing both kinds of body with active forces of motion as well as with passive forces of inertia and impenetrability. Leibniz's inorganic bodies, lacking dominant entelechies, cannot be regarded as actual agents generating their own spontaneously initiated motion, but we have seen how they might still be regarded as having some kind of causal agency deriving from the spontaneous motion and resistance of their organic parts.[46] Just as each part of an inorganic body moves and resists by means of its own derivative active and passive forces, so too the inanimate aggregation of these parts may be said to move and resist, not by the imposition of an external force, but solely by the composite activity and passivity of its internal parts. Hence, for Leibniz, each billiard ball – to take the classic example of inanimate aggregate body – moves by its own derivative active force at the exact moment when the colliding ball loses an equivalent amount of active force, giving the overall appearance that force has been transferred from one ball to the other. In this sense, Leibniz's nonliving objects have power that is not unlike the nonliving agentic power envisaged by Bennett: they move and resist independently of any external intervention from human or other living beings, their motion resulting entirely from the spontaneous motion of their organic parts.

The agency and motion of the parts also helps explain the porosity that Bennett highlights as a characteristic of nonliving matter, for it is due to the self-motion of bodies' parts that there is constant flux and mutual exchange of parts across different bodies. For Leibniz, all bodies are porous to varying degrees, like sponges that take up matter from outside and that, in the less dense bodies, 'glides through their pores' (*On Body and Force*: GP IV 395; AG 252). The boundaries are always fluid. We have seen how in living things the biological processes of ingestion and excretion effect the mechanical exchange between the inner and the outer,[47] but inanimate things too have fluid boundaries. Nonliving things do have sufficient continuity of structure to allow us to identify them as the same objects (or events) over time. Some, like mountains, planets or stars endure over very long time spans; others, like ice or sand artworks, rainbows or sunsets, persist for much shorter durations. Nevertheless, as with living bodies, so too for inanimate bodies, continuity of structure, for however long it lasts, does not prohibit the constant interchange of parts at the boundary between a body and the external environment. Pebbles, stones, rocks and mountains are subject to erosion and weathering; clouds form, change shape and dissipate; plastic toys, bags and other man-made artefacts eventually photo-degrade; soil, wood and cloth soak up water; carpets and furnishings fade in sunlight, reflecting changes in atomic structure at the object–environment interface. Indeed, nothing remains the same internally or externally from one moment to the next.

For the most part, the changes are incremental and imperceptible. Glaciers do occasionally move at great speed and in the past century they have been retreating at an unprecedented rate, but until then glaciers, initially formed in the last Ice Age, were relatively stable, slow-moving structures that illustrate the slowness and the 'patience, stability, duration' that Bennett discerns in inanimate things (Ja. Bennett 2012). Typically, insentient physical objects, whatever their size, change or move only gradually. Objects remain where we place them. We find things in the morning much as we left them the night before. Bodies' impenetrability and inertia are the physical manifestations of their derivative passive force, which in turn results from the modification of the primitive passive forces in their monadic constituents (*Specimen Dynamicum*: GM VI 236–237; L 437). Slow-moving or incrementally changing material masses, we may presume, are aggregates of substances whose dominant monads have an abundance of primitive passive force or primary matter. These are the living things, hidden in the infinitesimally small portions of the material continuum, that have not yet been roused into activity, but that are, as it were 'asleep in the abyss of things' (*On the Ultimate Origination of Things*: GP VII 308; AG 155), whose perceptions are insensible and confused and whose bodies mirror this confusion in their overabundance of resistance to change.

In the passive materiality of inanimate things and in our own corporeality, Jane Bennett finds the reason for our human sympathy with material things. She believes that we are able to sympathize with the inorganic because we ourselves possess an inorganic nature – the fact, as she puts it, that we are all 'stuff'. Leibniz too believes that we experience a kind of 'sympathy' with inanimate things when we passively or confusedly sense perceive their innate perfection without consciously or intellectually recognizing it as such. The feelings of pleasure that we experience when we sense the perfections of things in this way foster a sympathetic engagement with them.

> We do not always observe wherein the perfection of pleasing things consists, or what kind of perfection without ourselves they serve, yet our feelings [*Gemüth*] perceive it, even though our understanding does not. We commonly say, 'There is something, I know not what, that pleases me in the matter'. This we call 'sympathy'.
>
> (*On Wisdom*: GP VII 86; L 425)[48]

We experience pleasant sensations or feelings in the presence of perfect and beautiful things. Their beauty affects us even though we do not distinctly and consciously perceive the full extent of the complex diversity, order and harmony in which their beauty resides. We sense their beauty and perfections and feel pleased, although our confused sense perceptions do not enable us to comprehend *why* we feel the way that we do or to explain to others what it is about the object that appeals to us. This sense of pleasure is, for Leibniz, a sympathetic response. We might hope that it encourages us to 'tread lightly' in the presence of the beautiful.

Bennett thinks that our sympathy for the inanimate results for the passivity we share with material things. For Bennett, such sympathy is tied to our passivity. To some extent this would seem to be true for Leibniz as well. After all, it is when we passively and confusedly sense perceive beautiful objects that sympathy arises. However, from a Leibnizian perspective, it is perhaps our active intellectual engagement with the other that is the more significant for the development of ecologically positive feelings towards both living and nonliving things. Although we share passivity and materiality with all things, we also share their active agency. And it is when we actively engage with them on both a physical and an intellectual level that we develop feelings, not merely of sympathy, but of love and in respect of living things, feelings of empathy (chapter 7). Leibniz holds that true love is directed only towards living things,[49] but feelings akin to love can be directed towards inanimate things:

> a painting of Raphael affects him who understands it, even if it offers no material gains, so that he keeps it in his sight and takes delight in it, in a kind of image of love.
> (*Codex Juris Gentium Diplomaticus*, preface: GP III 387; L 422)

Leibniz's claim that Raphael's paintings need to be not only sense perceived but also understood serves to remind us that the true appreciation of beauty, the kind of appreciation that generates an 'image of love', demands some effort on our part. It requires us to pay close attention to the things themselves and to perceive them as distinctly as it is possible for a finite creature to do. We perceive material things through the medium of our own bodies and organs of sense. Such perceptions always involve a degree of confusion,[50] but we can make them less so. We cannot individually identify the infinitely small moving (and living) parts of objects, but we can be alert to the finer details of the visible, auditory, tangible and other sensory qualities of objects and can invoke our intellectual understanding of perfection and harmony in order to recognize and appreciate their beautifully unified order and variety.

In our sense perceptions of living things, attention to detail enables us to fully recognize each individual as a distinct being in its own right. Imagine that we see a couple of birds in the air. If we can be sure that they are blue tits rather than coal tits, then, following the epistemological framework set out in Leibniz's *Meditations on Knowledge, Truth, and Ideas*,[51] our perception may be said to be both clear (for we can distinguish the birds as things distinct from their surroundings) and distinct (for we can distinguish them well enough to know that they are not coal tits). In this case, we pick out the distinguishing features that mark out the birds as members of their species. Were we also to notice the smaller differences between the blue tits themselves – such as one being larger, more aggressive, more brightly coloured than the other – our sense perceptions of them would be even more distinct, for we would then be able to identify each not only as a blue tit, but as this *particular* blue tit, and would be able to reidentify each one as that particular blue tit on encountering them again in the future. The more we clearly and distinctly perceive the characteristics of each of the birds, the more we fully appreciate each one as the unique individual living being that it is. And the more we do so, the more we realize that each bird not only stands in relationship to ourselves, but that each is also, like ourselves, an independent being with its own distinct identity and unique perspective on the world.

To appreciate this fact, we must first become self-consciously aware of ourselves as perceiving beings, distinct from the things that we perceive. We are then in a position to realize that the things we perceive are correspondingly distinct from ourselves. We are able to appreciate the bird we perceive as an independent unique entity, with its own way of being in the world, its own distinctive features, its own perceptions and perspective on the world, its own life and interests. We become aware of it as a being in its own right and come to realize that there is indeed something that 'it is like' internally or experientially to be this particular bird and that that something is quite different from what 'it is like' to be our-selves. The 'what it is like' to be this bird is this bird's own perspective or point of view: the point from which it perceives the world, focusing with different degrees of clarity on items that interest it (the bird, for instance, will be far more aware of the movement of a worm, or of the person distributing bread-crumbs than we might be). In this way, the distinctness of our own perceptions, encouraged by our attentiveness to detail, leads to our awareness that it is we who perceive this bird, that we perceive this bird rather than any other bird, and that this bird, in all its particularity exists – as both perceiver and perceived – in relation to us. Our awareness of this relationship cannot help but foster in us a further sense of responsibility and care for this unique individual with whom we coexist.

Although inanimate things cannot be said to experience the world from their own distinct 'points of view', this does not mean that they do not also warrant our attention. Paying attention to the things around us is of utmost importance in the development of an ecological sensibility and this is as true for the living as for the nonliving. Attending closely to natural phenomena such as cloud or rock formations, landscapes, rainbows, sunsets, grains of sand and clods of earth, being keenly aware of the sounds and smells around us, of the heat of the sun or the cool of the mountain spring, makes us aware of the incredible diversity or plurality and the harmony and order within the natural world and exposes its remarkable beauty.

Leibniz defines beauty as that 'the contemplation of which is pleasant' (*Elements of Natural Law*: A VI i 464; L 137). This does not mean that beauty is a subjective quality that resides only in the beholder. Rather, the beholder experiences pleasure when contemplating the beauty that is already present as an objective quality in the thing contemplated. Beauty exists independently of whether it is perceived by us or not. But when we do become aware of it, we feel a sense of pleasure. Such pleasurable contemplation of beauty is disinterested insofar as it is not the kind of pleasure we might feel from finding something useful. The love of God, i.e. the love of the most perfect being that is the source of our greatest pleasure, is a disinterested love in this sense (*Principles of Nature and of Grace* §18: GP VI 606; AG 212). The love of God is manifested in the love of God's creation.[52] The love of individual finite beings and things that is generated by the pleasurable contemplation of their beauty and perfection is similarly disinterested and disregarding of any utility they might have for us. The pleasure we gain from the contemplation of beautiful things is not the kind of pleasure that Descartes confesses to feeling at the sight of the ships entering harbour laden with goods.[53] Were we to take pleasure from the sight of the beauty of the ships, we would do so on account of, for instance, the sheer ingenuity of their construction and the grace and ease with which they sail through the water and into the harbour.

The more that we understand nature and the living and nonliving things within it, the more that we learn of their structure and how their parts combine, especially when we understand this through scientific investigation, then the more fully we experience nature's beauty:

> whenever we penetrate to the basis of anything, we find there the most beautiful order we can desire, surpassing anything we had expected, as anyone knows who has understood the sciences.
> (*On What Is Independent of Sense and of Matter*: G VI 507; L 552)[54]

Telescopes reveal the 'spectacular' structures of the stars, but Leibniz favours the microscope, partly because discoveries in this field are useful in furthering human health and wellbeing, but also because they are more successful in uncovering the beautiful internal structures of things (*Reflections on the Common Concept of Justice*: L 566). The microscope discloses far more of the minute details of living things than even the closest of unaided visual inspection could ever achieve. However, whether with the aid of a microscope or simply with the naked eye and other senses, attending closely to the objects of human design also alerts us to their inherent beauty.

Beauty is not the sole preserve of natural things. It can be found too in 'lifeless creation, a painting or a work of craftsmanship' (*On Wisdom*: G VII 86; L 425). Indeed, beauty can be discovered just as much in humble, everyday things, such as a wholesome meal, a comfortable sofa, a well-fitting suit or well-designed pen or spoon[55] as it can in fine works of art, architectural treasures or remarkable feats of engineering. All can possess the beauty that arises from the harmonious ordering and coming together of a multiplicity of diverse parts. Through deliberate, conscious attention, we are made aware of their perfection and beauty. In the active contemplation of such beauty, we feel a sense of pleasure. It is not unreasonable to suppose that in the realization of the importance that the sense of beauty plays in our lives, we will also desire to protect and preserve it, and, wherever possible contribute towards it.

It was Leibniz's firm belief that the universe itself is the supreme example of the perfectly balanced harmony of order and variety that constitutes beauty: in the universe, everything is 'regular and rich beyond what anyone has previously conceived; with matter everywhere organic – nothing empty, sterile, idle – nothing too uniform, everything varied but orderly' (*New Essays*: A VI vi 72–73; RB 72–73). The beauty of the whole, however, is not apparent at each moment. Instead, it unfolds over time. Because each entelechy or soul mirrors the whole universe, albeit obscurely, Leibniz contends that '[o]ne could know the beauty of the universe in each soul, if one could unfold all its folds, which only open perceptibly with time' (*Principles of Nature and of Grace* §13: GP VI 604; AG 211). Leibniz likens the unfolding of the perfection of the universe to the unfolding of a novel – its beautiful composition would be less perfect if 'the reader could divine the entire issue at once'. The 'beauty of a novel' lies in the way in which 'order emerges from very great apparent confusion'. We would not wish to 'take a novel by the tail and to claim to have deciphered the plot from the first book'. The universe itself is like a 'great and true poem', a 'word-by-word creation' whose perfection is not revealed to us all at once (*Reflections on the Common Concept of Justice*: L 565–566).

The full perfection of the world is in process of unfolding, but our appreciation of beauty in the present sets the tone for the future, in the same way as our present 'love of God already gives us, here and now, a foretaste of future felicity', making us content in the present moment while also assuring us of 'future happiness'. The love of God – and we might add the contemplation of the beauty of God and of God's creation – not only 'fulfils our hopes', but also,

> leads us in the way of supreme happiness, since, by virtue of the perfect order established in the universe, everything is done in the best possible way, as much for the general good as for the greatest particular good of those who are convinced of it and are satisfied by the divine government. This cannot fail to be true of those who know how to love the source of all good.
>
> (*Principles of Nature and of Grace* §18: GP VI 606; L 641)

In the perception and contemplation of beauty we find a sure guide for our actions. Our attentive awareness leads us to the appreciation of the beauty that exists all around us, in the living and the nonliving, in the natural and the artificial. Recognizing the beauty of divine creation in the present moment, we cannot help but will to play our part in the unfolding of greater perfection in the future, aiming to increase diversity while retaining orderliness as far as we can, striving for harmony, variety and order, balance and unity in all we do, and taking pleasure in perfection, not because it is useful to us, but simply because it is beautiful, or because it is a part of the beautiful whole, the perfect creation of a perfect God. Faced with the choice of building on a piece of derelict wasteland or on a site of special scientific interest or great natural beauty, the principle of perfection (or beauty) favours bringing the derelict land back to life, restoring it to a place in which as many forms of life as possible can flourish and harmony prevail. And because beauty, as an objective quality, is present even when we do not consciously perceive it, we need to be mindful to protect those things and places hidden from our immediate view. Were we to factor rationalist aesthetic considerations of beauty into our moral, social or political deliberations, we might perhaps be less inclined to clutter the awe-inspiring vastness of space with our debris or to hide our waste in the depths of the oceans.

In this chapter, we have explored some reasons for regarding both the organic and inorganic as valuable and worthy of ethical and aesthetic consideration. In the next, we consider the value of things in terms of the relations that obtain between them and in terms of their relation to the world as a whole.

Notes

1 Chapter 2, p. 38.
2 See Phemister 2005, chs 2 and 3.
3 Smith 2011, 72–73. For a useful survey of early modern theories of nutrition, see Smith 2011, 74–78.
4 These features, however, cannot conclusively delineate the boundary between living and nonliving things. See Phemister 2011.
5 See also: 'in the regular assemblages of nature, that is, in the organized bodies like those of animals, there are dominant unities whose perceptions represent the whole; and these unities are what are called "souls", or what each person means when he says "I"' (to Sophie, 19 November 1701: A I xx 75; LS 209).
6 But see chapter 2, note 44, p. 48. We shall also discover in chapter 8 that the body requires that its soul contain all the traces of the past and lineaments of the future in order that the body too may retain all past effects upon it, a condition that Leibniz considers necessary if the body is to enter into (efficient) causal relations with other bodies (chapter 8, note 15 (p. 176)).
7 Leibniz rejected the immaterial vital principles and plastic natures postulated by the Cambridge Platonists. He thought there was no need to assume immaterial forces as immediate causes of the motion and organization of bodies: *material* plastic natures (derivative physical forces) could perform that role just as effectively without invoking seemingly inexplicable interactions between the physical and the nonphysical. For discussion, see Smith and Phemister (2007).
8 Kant favoured epigenesis, the view that new forms or organized bodies arise naturally from existing ones (*Critique of the Power of Judgment*: AK V 422–423; CPJ 291).
9 See also, *Considerations on Vital Principles and Plastic Natures*: GP VI 544; L 589.
10 See e.g. *Monadology* §74: GP VI 619–620; AG 222.
11 It should be noted here that for Leibniz, no indivisible living thing or monad can be destroyed by natural means. See, for instance, *Monadology* §§4–6 (GP VI 607; AG 213) and *Discourse on Metaphysics* §9 (GP IV 433–434; AG 42). What we consider as death consists only in a sudden transformation in which the core of the creature's body is contracted to an invisible physical point and its perceptions reduced to insensible confusion. See chapter 3, note 11 (p. 63). Leibniz believed that the fate of humans is rather different. See below, note 13. We explore some implications of creatures' indestructibility in chapter 8.
12 In this respect, the smaller animals are just like the larger ones. They too 'grow from other smaller, spermatic animals, in proportion to which they may be considered

large; for', Leibniz insists, 'everything goes to infinity in nature' (*Principles of Nature and of Grace* §6: GP VI 601; AG 209).

13 Opinion is divided as to whether Leibniz actually believed that individuals might species-jump in this way. On the affirmative side, he does maintain that humans after death will have 'subtle bodies' of a different and more angelic form than their gross human ones (*Reflections on the Doctrine of a Single Universal Spirit*: GP VI 533; L 556–557), which suggests that they will be more angelic than human, and in stating at *Monadology* §75 that the majority of animals 'remain among those of their kind' (GP VI 620; AG 222), he implies that there are some animals that do not always 'remain among those of their kind'. Moreover, Leibniz cautiously acknowledges early theories of the evolution of animals and gives the impression that he would accept them were it not that they go against 'the sacred writers, with whom it is impious to disagree' (*Protogaea* §6: Leibniz 2008b, 14–15). His youthful interest in alchemical processes might well also have encouraged him in the belief that individual living things need not always belong to the same species as they do now. On the negative side, however, Leibniz maintains against Locke that generation gives a 'strong presumption' of species membership even in the face of seemingly 'monstrous' births (*New Essays*: A VI vi 315; RB 315). The implication is that individuals remain species-bound with the seeds of future humans residing in present human beings, the seeds of future dogs residing in present dogs, and so forth, even though sometimes the shape of the body belies the source. Nevertheless, even here Leibniz admits there is insufficient evidence to decide the issue, instancing tales of hybrid plants and animals that have the reproductive capacities to perpetuate the new species (*New Essays*: A VI vi 315; RB 315).

14 In the midst of these changes, some so radical that they entail species change, can we still identify a 'thing' that retains its identity through change? Leibniz thought so, at least in the case of living things. These not only have constantly changing bodies; they also have dominant entelechies or souls that unify the body and make the whole animal or animate being into an indivisible and indestructible substantial thing. All the changes in the animal, both psychological and physical, are governed by its dominant entelechy's unique law or essence, an essence that unfolds as the ordered sequence of perceptions and appetitions in the soul and as the changing states of its body, the body having been preformed or 'regulated' by God so that their mechanical interactions are guaranteed to produce the exact physical correlates to the soul's psychological states. In his Fifth Letter to Clarke, Leibniz maintains as much for the rational or free soul (§92: GP VII 412; LC 85–86), but the same must hold for all living bodies and their dominant entelechies, all the way down.

15 Tom Bristow's discussion of Foer's novel is recommended (Bristow 2012).

16 If some of these are future humans, we might even be prone to cannibalism. This was a live issue in Leibniz's day, arising in the course of discussion about the resurrection of the body after death. See Strickland 2009 and 2010.

17 This gives a novel and idiosyncratic twist to the common translation of Leibniz's claim in the *Monadology* that every simple substance is 'pregnant [*gros*] with the future' (§22: GP VI 610; AG 216). Temporal dimensions of substances are discussed in chapter 8.

18 See Phemister 1999 and 2005, chs 2 and 3.

19 On the nested structures of Leibniz's organic bodies, see Nachtomy 2007, chs 9 and 10.

20 On his return to Hannover from Paris, Leibniz had sought out leading microscopists Jan Swammerdam in Amsterdam and Anton van Leeuwenhoek in Delft (Antognazza 2009, 177). For a fascinating account of microscopy in early modern science and the philosophical interpretations of its results, see Wilson 1995a.

21 There is therefore no metempsychosis but only metamorphosis, for 'souls never entirely leave their body, and do not pass from one body into another that is entirely new to them' (*Principles of Nature and of Grace* §6: GP VI 601; AG 209).

22 *Considerations on Vital Principles and Plastic Natures* (GP VI 543; L 589); *New Essays* (A VI vi 328; RB 328).

23 See also note 11, p. 89 above.

24 Strickland translates *échantillons architectoniques* as 'smaller-scale constructions'. Ariew and Garber (AG 223) translate the phrase as 'schematic representations'. Strickland's translation better captures Leibniz's meaning. For discussion see M 150–151.

25 When it does not, however, Leibniz's optimism leads him to believe nevertheless that 'the apparent deformities of our little worlds combine to become beauties in the great world, and have nothing in them which is opposed to the oneness of an infinitely perfect universal principle: on the contrary, they increase our wonder at the wisdom of him who makes evil serve the greater good' (*Theodicy* §147: GP VI 198; H 216).

26 This gives us all the more reason to heed the warning in a 2005 European Communities report on synthetic biology that it is 'important to address ethical and safety concerns, and to address potential or perceived risks of synthetic biology from the very beginning' (European Communities 2005, 5). The same report worryingly and somewhat hubristically suggests that nature can be improved by human intervention when it introduces synthetic biology as a science that 'will enable the design of "biological systems" in a rational and systematic way' (European Communities 2005, 5). Leibniz would surely have pointed out that the divine creation – the best possible world – is already 'rational and systematic'.

27 The point has been recognized by Paul Rateau, who also cites in this connection the opinion of Jean Baruzi (Rateau 2008, 291).

28 Commentators are divided as to whether all Leibnizian creatures would be perfected given an infinite period of time and Leibniz's own opinion on this issue fluctuated. In the mid- to late 1690s, he suggested that there were some creatures whose advancement has yet to begin (*On the Ultimate Origination of Things*: GP VII 308; AG 155), which implies that, given sufficient time, it will begin at some point. However, in 1712, Leibniz admitted to Louis Bourguet that some seeds never develop into mature plants (to Bourguet, October 1712: GP III 559). Of course, this may be because their development takes a different direction, but equally, it may be because Leibniz did not by this time believe that the perfecting of things proceeded indefinitely in all creatures. We return to the issue of the perfectibility of creatures in chapter 8.

29 The perfections of the soul and the perfections of the body are examined in chapter 7.

30 See also: 'each soul is a mirror of the universe in its own way, without any interruption, and contains in its depths an order corresponding to that of the universe itself; and that the souls vary and represent in an infinite number of ways, all different and all true, and thus multiply the universe, so to speak, as often as possible, and in such a way that they approach divinity as far as they can in their different degrees and give to the universe all the perfection of which it is capable' (*Reflections on the Doctrine of a Single Universal Spirit*: GP VI 538; L 559–560).

31 *Theodicy* §124: GP VI 178–179; H 198.

32 These reasons to value humans should presumably apply with even greater force to the valuation of angels. If others are considered lesser beings than humans, then correspondingly, humans should be considered of less value than angels.

33 Smith 2011, 48–57. See also 2007, 144.

34 See also Sessions 1977, 486.

35 Smith 2011, 319n99 and Antognazza 2009, 463.

36 As we shall see in chapter 6, Leibniz defines happiness as 'lasting joy', and the steps towards true happiness are the pleasurable feelings we experience whenever we perceive beauty, virtue and other perfections in ourselves or in others. Thus, we come full circle: our happiness is grounded in our perceptions of perfection and most of all in the contemplation of the perfection of God and God's creation.

37 Nature will still serve us, but as an inflicter of punishment rather than reward. The harmony of nature and grace 'leads things to grace through the very paths of nature.

For example, this globe must be destroyed and restored by natural means at such times as the governing of minds requires it, for the punishment of some and the reward of others' (*Monadology* §88; GP VI 622; AG 224).

38 Indeed, under Leibnizian pluralism, the world as a whole is an inanimate aggregate that risks being undervalued in the absence of good reason to appreciate nonliving things. Naess's biospherical egalitarianism, grounded in Spinoza's absolute monism, avoids this problem.

39 The 'in principle' qualification is in order here. Just as deep ecology promotes 'biospherical-egalitarianism-in-principle', thereby allowing a degree of anthropocentrism when cases require, so too, 'ontoegalitarianism-in-principle' must allow room for biocentric preferences to take priority when needed.

40 Some commentators (e.g. Adams 1994, 246) maintain that the identity of bodies, organic and inorganic, is entirely mind-dependent. Others (e.g. Lodge 2001) argue only that there is a mind-dependent component to the identity of physical aggregates. Leibniz himself admitted mind-independent features of bodies by which their identities momentarily and over time can be sustained in the face of constant flux. He speaks of the parts of bodies running in the same direction. This allows that the internal relations between the parts of bodies can remain relatively stable while their relations to external bodies change – as happens, for instance, when a ball is thrown into the air. The parts of the ball move together, their relations to each other remaining the same while their relations to those in the rest of the environment change. See Phemister 2005, 124–125 and discussion of places, chapter 6, pp. 114, 124–125.

41 See Phemister 2005, 169–175.

42 Hence, a mere aggregate body – a house for instance – is a divisible thing whose apparent unity is destroyed by the separation of its parts.

43 Above, p. 75.

44 Leibniz's Fourth Letter to Clarke: GP VII 372; LC 36.

45 Ja. Bennett 2010, 9. Some of these actants are objects in the traditional sense, but Bennett also includes in her understanding of vibrant matter things like the North American electrical grid (Ja. Bennett 2010, 24–28).

46 Bennett herself notes that '[a]n assemblage owes its agentic capacity to the vitality of the materialities that constitute it' (Ja. Bennett 2010, 34).

47 Above, pp. 70–71.

48 Leibniz does not distinguish here the inanimate from the animate, so we may assume he intended this 'sympathy' to apply as much to the former as to the latter.

49 More precisely, in Leibniz's view, true love is directed only towards living beings capable of happiness. In chapter 6, we shall argue for the removal of this restriction.

50 See chapter 2, pp. 37–38.

51 GP IV 422–426; AG 23–27.

52 See chapter 6, p. 121.

53 Chapter 1, p. 13.

54 The view suggested here is not unlike the positive aesthetic theory of nature promoted by Allen Carlson (1984 and 2000, ch. 6). See also Carlson 2004, 71–72. For further discussion of Leibniz and Carlson on this matter, see Phemister and Strickland 2015.

55 Saito (2007) makes a persuasive case for the importance of aesthetic appreciation of everyday things.

5 Relationality and value

[*Philalethes*]: ... a change of relation can occur without there having been any change in the subject: Titius, 'whom I consider to day as a father, ceases to be so to morrow, only by the death of his son, without any alteration made in himself.'

[*Theophilus*]: That can very well be said if we are guided by the things of which we are aware; but in metaphysical strictness there is no wholly extrinsic denomination (*denominatio pure extrinseca*), because of the real connections amongst all things.

(*New Essays*, 2.25.5: RB 227)

The publication in 1992 of John O'Neill's article on intrinsic value brought much needed clarity to the philosophical confusion that until then had surrounded the concept (O'Neill 1992). O'Neill identified three distinct senses and uses of the term 'intrinsic value'. The first defines 'intrinsic value' negatively as an absence of instrumental value. A thing may be regarded as having non-instrumental intrinsic value when the value ascribed to it can be stated without reference to any instrumental value that the thing may have in furthering the needs and wishes of human beings or in furthering the needs and wishes of any other sentient being or group of beings. Intrinsic value is not to be ascribed to hammers or to sticks on the basis that they have instrumental value for humans in the construction of buildings or furniture or instrumental value for the crow in clearing the ground to expose bugs or worms. If the hammer and the stick, or the human and the crow, do have intrinsic value, this has to be justified by factors or reasons other than any use-value they possess in relation to others.

As O'Neill recognizes, this first sense of 'intrinsic value' is central to ethical environmentalism: 'To hold an environmental ethic is to hold that non-human beings have intrinsic value in the first sense: it is to hold that non-human beings are not simply of value as a means to human ends' (O'Neill 1992, 120). In this context, O'Neill cites Naess's claim that '[t]he well-being of non-human life on Earth has value in itself. This value is independent of any instrumental usefulness for limited human purposes' (O'Neill 1992, 119). Nevertheless, this first sense only tells us what intrinsic value is not; it does not tell us what it is. O'Neill is right therefore to suggest that a 'defensible ethical position about the environment' may also require a commitment to either the second or the third sense of 'intrinsic value' (O'Neill 1992, 120).

O'Neill's second sense maintains that a thing's intrinsic value is grounded in those of its intrinsic or internal properties that are value-conferring properties. They are the kinds of properties mentioned in the previous chapter, properties such as life, beauty or the capacity to flourish. The term 'intrinsic' is used here to refer to properties that a thing possesses in itself, independently of any relations the thing has to external things. Properties that are intrinsic in this sense do not necessarily confer value, although some do. The key point is that intrinsic properties are here regarded as essential on the ground that any changes to the intrinsic or internal properties of a thing are considered to constitute changes to the thing itself. Changes to the intrinsic properties constitute changes to the essence or nature of a thing. It is further commonly assumed, following G. E. Moore, that intrinsic properties are nonrelational properties. In other words, they do not depend upon, nor are they affected by, relationships that a thing may have to external things. Extrinsic – or external – properties, in contrast, are properties a thing possesses insofar as it does stand in relation to external things. Changes in these external relationships are accompanied by corresponding changes to the thing's extrinsic properties, but its intrinsic properties remain intact. That is to say, in the manner outlined by Philalethes (representing Locke) in the quotation at the beginning of this chapter, changes in the extrinsic relations that a thing has to external things are not thought to incur any real changes in the things themselves. Extrinsic properties are therefore regarded as nonessential properties insofar as changes to these properties are not considered to alter the thing's essence or nature. Spatial relations between physical objects provide the standard example of extrinsic or external properties. For instance, the spatial location of an orange is a nonessential extrinsic property. The orange is the same orange (it has the same essential intrinsic properties) regardless of whether it is located on the shop shelf or in the fruit bowl at home. Similarly, a book remains the same book, with the same intrinsic properties, even when its extrinsic properties are changed by moving it from the bookcase to the table.

O'Neill's third sense of 'intrinsic value' considers intrinsic value as 'objective value': value that is 'objectively real' or actually present in the thing itself independently of any subjective evaluation by an external observer. Something possesses extrinsic value only if an external valuing subject, such as God or a human being, has subjectively deemed its properties to be valuable. The thing itself is presumed to have no value until or unless such a value judgement has been made. Intrinsic values, in contrast, belong to the object in itself. They exist in things themselves independently of any observer: they are discoverable, but not necessarily already discovered. Intrinsic values do not need to be subjectively imposed by an external agent; they do not depend upon any external observer's judgement. For instance, on Leibniz's view, beauty is objectively valuable. The world that God created is a beautiful world, but insofar as the world is objectively beautiful, its beauty is wholly independent of God's will. The world is beautiful in itself. God chose to create this world *because* it is the best; it is not the best *because* God chose it.

Intrinsic relationality

In this section, I take issue with the idea of intrinsic properties as nonrelational properties, as employed in the second of O'Neill's senses of intrinsic value. In reporting on actual usage of the term, O'Neill had no reason to question Moore's view that intrinsic properties are always nonrelational. However, it will be argued here that, from a Leibnizian perspective, intrinsic or internal properties may be both essential *and* relational. After this, we return to questions about value, exploring ways in which we might re-envisage and revise our understanding of intrinsic internal properties not only as relational, but also, in each case, as value-conferring properties.

Moore's correlation of the extrinsic with the relational and the intrinsic with the nonrelational underpins Bertrand Russell's reading of Leibniz's doctrine of the complete concepts of individual substances. Assuming a distinction between intrinsic, nonrelational monadic predicates and extrinsic relational nonmonadic predicates, Russell contends that only the former are included in an individual substance's complete concept. Relational nonmonadic predicates arise subsequently only through the comparison and contrast between the complete concepts of diverse substances. Ultimately, then, relational predicates are grounded in, and reducible to, nonrelational monadic predicates:

> According to Russell, Leibniz's commitment to a subject/predicate logic or, more precisely, to an ontology which is described solely by means of monadic predicates commits him to reduce statements with relational predicates to statements with monadic predicates.
>
> (Nachtomy 2007, 90)

Monadic predicates describe the monads' perceptions. These, Russell holds, are the nonrelational foundations of the relations that God understands to hold between substances (Russell 1992, 14). God knows what any one substance will perceive if he chooses to create it and knows too which other substances will harmonize with these perceptions if he chooses to create them as well. In this way, relations – although true because they are grounded in the 'perceptions of phenomena in simple substances' are nevertheless only 'the work of the mind' (Russell 1992, 14). However, the relations themselves, as purely external, make no difference to the natures of the individual substances whose essences are encapsulated in their entirety by the internal, intrinsic, nonrelational monadic predicates of their complete concepts.

In support of his interpretation of Leibnizian relations as external and ideal, Russell cites the passage from Leibniz's Fifth Letter to Samuel Clarke in which Leibniz describes three ways of considering the ratio between two lines, referred to as 'L' and 'M'.[1]

> The ratio or proportion between two lines L and M, may be conceived three several ways; as a ratio of the greater L, to the lesser M; as a ratio of

the lesser M, to the greater L; and lastly, as something abstracted from both, that is, as the ratio between L and M, without considering which is the antecedent, or which the consequent; which the subject, and which the object. ...

In the first way of considering them, L the greater; in the second, M the lesser, is the subject of the accident, which philosophers call relation. But which of them will be the subject, in the third way of considering them? It cannot be said that both of them, L and M together, are the subject of such an accident; for if so, we should have an accident in two subjects, with one leg in one, and the other in the other; which is contrary to the notion of accidents. Therefore we must say, that this relation, in this third way of considering it, is indeed out of the subjects; but being neither a substance, nor an accident, it must be a mere ideal thing.

(GP VII 401; LC 71)

The third way of conceiving the ratio between the two lines regards the relation as abstracted from the lines themselves. This relation must be an external relation because otherwise, as Leibniz explains, each line would possess the same accidental or inessential quality. Since qualities cannot be shared in this manner, the relation of the lines to each other must be purely external to both of the relata. Russell is correct to claim that Leibniz regards such external relations as abstract or ideal. They exist between things, but do not belong to the things themselves as internal properties.

However, Russell also maintains that even in the first two ways of conceiving the relation between the lines – where the subject of the relation is either L or M but not both – each of the subjects must be specified solely in terms of its intrinsic, nonrelational predicates before it can be a subject in the relation. The lines, L and M, must first be conceived independently of each other, with the respective relations of 'greater than' or 'lesser than' only coming into being when the two lines are compared with each other. Russell's view makes sense intuitively: two things can only stand in relation to each other if they are indeed two distinct and separable things. However, Russell's distinction between nonrelational, monadic predicates and relational, but nonmonadic predicates presumes that each substance must itself be nonrelational in the sense that *all* of its predicates are so. However, it surely suffices that *some* of the predicates in the substance's complete concept are – or were originally – nonrelational. It is not necessary that monadic predicates in the individual's complete concept exclude all references to its relations to other individuals.[2]

More recent interpretations have highlighted a number of passages in which Leibniz emphasizes the presence of relational predicates in the characterization of individual substances. Even in the case of the two lines, L and M, it is possible that the predicates that ground the ideal relation between them are relational predicates, as for instance, that L as subject contains the relational predicate of being longer than M and conversely that M as subject contains the relational predicate of being shorter than L. Hidé Ishiguro was one of the first to realize that for Leibniz, 'there is no way of characterizing things without invoking both the relational properties and the non-relational properties of the things in question' (Ishiguro 1990 [1972], 107).[3] Leibniz himself is quite explicit about the fact. In a letter to Arnauld, dated 4/14 July 1686, he states that,

> the concept of the individual substance contains all its events and all its denominations, even those that one commonly calls extrinsic (that is to say, that belong to it only by virtue of the general connexion of things and of the fact that it is an expression of the entire universe after its own manner.
>
> (GP II 56; LA 63)

Complete concepts of individuals do not contain only monadic nonrelational predicates, as Russell believed, but include also its so-called 'extrinsic denominations' that relate the substance (when created) to all the other created substances in the universe. In saying that 'extrinsic denominations' are only 'so-called', Leibniz implies that even these are properly internal or intrinsic predicates. All the predicates included in a monad's complete concept, including its relational predicates, are intrinsic. The so-called extrinsic relational denominations are actually *intrinsic* relational predicates. The predicates in the concepts of individual substances make reference to its relations to every other substance in the universe:

> there is nothing in the universe of created things that does not need the concept of every other thing in the universe for its perfect concept, since each thing influences every other in such a way that, if it were imagined that that thing were removed or different, everything in the world would be different from what it is now.
>
> (To de Volder, 6 July 1701: LV 208–209)

With infinite precision, each individual's complete or perfect concept includes the ways in which that substance relates to every other in the same world. *Intrinsic relational* properties are essential in characterizing each individual in all its particularity. Were even the slightest relation altered, the individual would be a different individual and would belong to a different possible world. When the substances are created, the relational predicates in their complete concepts are manifested as the internal relational qualities – or modifications – of the individual substances. These internal relational qualities then serve as the ground or reason for the extrinsic or external and ideal relations between things. So, for instance, the respective relational qualities in David (as father of Solomon) and Solomon (as son of David) ground the external abstract relation between them:

> paternity in David is one thing, filiation in Solomon another, but the relation common to both is a merely mental thing, whose foundation is the modifications of the individuals.
>
> (To Des Bosses, 21 April 1714: LB 326–327)

Hence, although the relations between any two or more created things are extrinsic, these external relations are grounded in relational qualities that are intrinsic to each individual in the relation. Caesar possesses the relational quality of being the father-in-law of Pompey. Conversely, Pompey possesses the relational property of being the son-in-law of Caesar. These qualities are relational insofar as they relate Caesar and Pompey to each other, but they are also intrinsic (internal and essential) insofar as both Caesar and Pompey would not be the people that they are if they were not so related. The relational predicate of being the father-in-law of Pompey is included in Caesar's complete concept in just the same way as the relational predicate of being the son of David is included in Solomon's complete concept.

These remarks apply universally. None of us would be the individual persons that we are did we not have the precise upbringing that we did, in determinate locations, making friends with particular people, forming bonds with individual pets, attending specific schools and being taught by certain teachers, and so forth. The contexts in which we live forge our individual identities. Our identities are constituted through our relations to others, no matter how far distant, in the same world and in part by the changes that occur in the others to whom we stand in relation: 'there are no extrinsic denominations, and no one becomes a widower in India by the death of his wife in Europe unless a real change occurs in him' (*On the Method of Distinguishing Real from Imaginary Phenomena*: GP VII 321–322; L 365).[4] Our complete concepts include all our relations to all other individuals in the same world: 'the notion of an individual is specified and determined precisely by its relations with other possible individuals and particular events' (Nachtomy 2007, 110).[5]

Nachtomy labels complete concepts of individuals as 'thick' or perfect concepts distinguishing them from 'thin' or incomplete concepts (Nachtomy 2007, 108). As incomplete, thin concepts are incapable of individuating substances located in a particular world. They need further refinement to make them thick or complete. Many incomplete concepts might contain the concept of kinghood, but only the complete concept of Alexander contains the concept of being *this* particular king with *these* particular subjects, both his and their lives described in such precise and minute detail that each completely specified individual can belong to one, and only to one, possible world. Nachtomy explains how in the pre-creation realm of logical possibility, the divine attributes – the absolutely simple forms – are combined and contrasted with each other so as to produce more complex concepts (Nachtomy 2007, 22–24). Further combinations and refinements lead to the formation of concepts that are so detailed and so fully interwoven with other detailed concepts that they form the complete concepts that contain all the relational predicates required to fully characterize each possible individual substance and locate it within the possible world made up of all the other individuals to which it will stand in real relation once it and they are created.

The process of filling out the incomplete concepts to form complete concepts of possible individuals and simultaneously forming the possible worlds to which these individuals belong is one of determining the infinity of possible relations between the incomplete concepts. The thin incomplete concept of 'Arnauld' need not contain any information as to whether this 'Arnauld' is married or not. However, the actual Arnauld in this world is an Arnauld who did not marry. In another – different – possible world, there may be another Arnauld (Arnauld*) who did marry and who thereby subsists in this realm of possibility within a web of different relations to other possible individuals, such as those of his wife's family, friends and colleagues. In this way, 'an individual is not fully individuated unless its relations to all other individuals are considered' (Nachtomy 2007, 111). Meanwhile, these other individuals are also in process of being 'fleshed out' so to speak. The incomplete concept of an 'Arnauld' who knows a similarly incompletely specified 'Malebranche' is not sufficient to determine the complete concept of the Arnauld who exists in this world. The complete concept of *this* Arnauld requires that the details of the 'Malebranche' with whom our Arnauld entered into public debate on the nature of ideas are also completed in such a way as to precisely identify the actual Malebranche, the author of the *Search after Truth*, with whom Arnauld would so vehemently battle intellectually. In this way, the identities of each unique, complete and possible individual are formed through its relations to others. As Nachtomy explains, 'relations to other (incomplete) individuals play a constitutive role both in forming worlds and in completing the individuality of possible individuals' (Nachtomy 2007, 110). The eventual completion is reciprocal: the concept of the Malebranche who subsequently exists in this world and whose concept includes reference to the Arnauld who subsequently also exists in the actual world is completed as much by its relation to the concept of Arnauld as the concept of Arnauld is in turn completed by the inclusion of predicates outlining Arnauld's relationship with Malebranche. On the creation of the actual world, the relationships established in the symbiotic formation of individuals' respective interwoven, co-constituted (even nondysfunctional codependent) identities are realized. Accordingly, monads' essences from which their perceptions and appetitions unfold and from which the derivative forces of their organic bodies derive are intricately bound up with the essences or identities of all other beings in the same world.

Faced with such deeply embedded interconnectedness and interdependence of substances, the question inevitably arises as to how we isolate and distinguish individuals from one another? The problem may appear intractable for the Leibnizian. After all, the particular specifications of every other being in the world are constitutive of any one being, for its relations to every other being are formative of its own identity. What entitles us to regard John and the blackbird, described in chapter 2,[6] as distinct and separate beings when their identities are characterized in terms of their relations to each other? The individual lives of living beings proceed as the passing succession of perceptions and appetitions, actualizing spatially and temporally the details contained in their complete concepts or, as Leibniz will later claim, harmoniously unfolding the 'laws of the series' that encapsulate each monad's essence or nature. The life of each is a succession of appetitions and perceptions, but none can be distinguished solely by the content of their perceptions, for each monad's perceptual state at any one moment is a mirror of the whole world, its past, its present and its future. Since all monads perceive or express the whole world at each moment, the description of each effectively describes the same world. If all monads have exactly the same perceptual content and each perception in each monad also has exactly the same content, how is one monad or even one perception in a monad to be distinguished from all the others?

Evidently, they are not distinguished by the content of their perceptual states. However, they can be distinguished one from another by the way this content is represented in each monad's perceptions. The relative degrees of confusion or distinctness in its perceptions determine the particular 'point of view' or perspective from which the monad expresses the world. No two monads have the same amount of active force either overall or from moment to moment. This primitive force, modified as the series of appetitions that propel the monad from one perception to the next, determines the degrees of distinctness of the monad's perceptions, thereby also establishing the monad's unique perspective or point of view. Our spatiotemporal location – the point from which we view the world – is parasitic on the relative clarity and distinctness of our perceptions: that which is perceived most clearly and distinctly is taken to be spatially and temporally close. We perceive our own bodies most distinctly of all. It is our own bodies, through which we perceive the world external to us, that are closest to us, temporally and spatially:

> since we perceive other bodies only through their relationship to ours, I was right to say that the soul expresses better what pertains to our body; therefore, the satellites of Saturn or Jupiter are known only in consequence of a movement which occurs in our eyes.
>
> (To Arnauld, 9 October 1687: GP II 113; LA 145)

Hence, even though the actual content of our perceptions is the same, the different perspectives from which this content is viewed, mediated through the perceptions of our own bodies, provides the means of distinguishing each substance as a unique individual.

Perceptions are internal or intrinsic qualities of monads.[7] On the Russell–Moorean assumption that only extrinsic qualities are relational, commentators have been encouraged to construe monads' internal qualities – their perceptions and appetitions – not only as intrinsic, but also as *non*relational. One prominent defence of this view emphasizes the internality and potentially solipsistic nature of Leibnizian perceptions. Appealing to Leibniz's claim in §17 of the *Discourse on Metaphysics* that 'each substance is a world apart, independent of everything outside of itself except God', Jan Cover and John O'Leary-Hawthorne contend that monads' perceptions are nonrelational qualities, for God could have created only one monad, the content of whose perceptions would in this case have no external referent. They argue that God is the cause of the pre-established harmony or correspondence between the various monads' perceptions and claim that relations arise only when God creates a plurality of monads, for only then do monads' perceptions correspond to, or come to be externally related to, perceptions had by the others (Cover and O'Leary-Hawthorne 1999, 75).

Cover and O'Leary-Hawthorne are right to assert that the actual correspondences or external relations come into being when God creates the plurality of monads whose perceptions harmonize with the perceptions had by all the others. God, we may say, by creating the plurality, establishes the ideal or abstract third kind of relations that Leibniz describes in his Fifth Letter to Clarke, the relation between the two lines L and M. However, that monads' perceptions can be externally related only if there exists a plurality of monads does not warrant the conclusion that a single monad's perceptions are entirely nonrelational. On the contrary, it is only because a monad's perceptions are internal relational qualities that the monad can be externally related when placed in conjunction with other monads and their perceptions. Although external relations are established by God through the creation of a plurality of monads, the metaphysical foundations of these external relations are the relational qualities internal to the monads themselves, namely their perceptions (and, as we shall see, their appetitions). On the basis of these internal relations, God foresaw the external relations in which substances would stand to each other were he to choose to create them. As internal relations, perceptions are like the first two of the relations described in the example of the lines L and M in Leibniz's Fifth Letter to Clarke, namely, the intrinsic relational properties of the lines themselves. Line L, as we saw, is greater than line M and conversely, line M is smaller than line L. Even in the realm of possibility, possible line L is greater than possible line M; even as only a possible line, L's concept includes the relational predicate of being greater than M. And even if God creates only line L without also creating line M, actual line L would still possess the relational quality of being greater than (possible) line M. Just as the complete concepts of possible individuals are co-constructed through the development of their relations with other possible individuals, so sizes and shapes are developed in relation to each other. To know of a possible line that it is six-inches long, we need to know that it is one-inch larger than five inches and shorter than seven; to know of a possible individual that it is Arnauld, we need to know, for instance, his relations to Malebranche and to Pierre Nicole, coauthor of the Port-Royal logic.

As British Hegelian F. H. Bradley recognized, all qualities must involve some kind of relation, whether they be qualities of perceivers or of physical objects. At the very least, they must involve difference relations that set the quality and the individual to which it belongs apart from others.[8] Take colour, for example. Blue is blue, but in being blue it is also not-yellow, not-red and so forth. Other relations also play a role in colours as experienced. Finely discriminated colours depend upon being seen in certain light conditions by living beings with the appropriate optical light receptors required to produce the appropriate sensation. The particular shade of blue in the curtain at this particular time of day requires that a whole host of relations hold – the precise way that the light falls through the window, the position of the sun in the sky that determines the angle that the light hits the curtain, the surrounding colours (the blue would look different placed next to red than if in relation to green) and so forth.

Even the traditional primary qualities, shape, size, motion, commonly thought to be the core intrinsic qualities that things possess independently of their (external) relations to other things, are actually fundamentally relational. The specific sizes and shapes of things in a plenum are as they are only because they stand in relation to others – the square top of the table could not be square were the things situated at its boundary placed differently. The precise size and shape of an object is constrained by what lies beyond its edges, even if this is only the whiteness of the paper on which a figure of a square may be drawn or the blackness of imagined empty space around a mental image of an imagined figure. Like shape and size, the primary quality of motion too would seem to be intrinsically relational. The motion of a football, for instance, is possible only if there is a space, populated by other objects against which the football appears to be moving as it changes its relations to them while they retain the same relations to each other.

Mental phenomena too are relational. Russell, followed by Cover and O'Leary-Hawthorne, had understood monads' perceptions as the nonrelational, intrinsic qualities upon which relations between monads are founded (Russell 1992, 14). As intrinsic or internal qualities, monads' perceptions and appetitions are constitutive of the monads' identities or essences.[9] Were a monad's perceptions not as they are, the monad would not be the monad that it is; it would instead be some other monad in some other possible world. In this sense, monads' intrinsic qualities are properties, i.e. essential qualities. But they are also relational qualities or properties, for monads' perceptions are also representations or expressions of the entire universe from the individual perceivers' unique points of view or perspectives. A soul or entelechy's perceptions have representational content that brings it into relation with all the other souls or entelechies and into relation with its own organic body and through this into relation with the bodies and the souls or entelechies of all other beings comprising this world.[10] Through their perceptions, all monads, whether minds, souls or entelechies, act like mirrors reflecting the whole,[11] each perception expressing the multiplicity of the actual universe in a single unitary experience. In the words of Ishiguro, 'although "… perceives" and "… perceives something" are monadic predicate expressions in the sense that they have only one blank space, they express relational properties' (Ishiguro 1990 [1972], 110). The same point is true also for monads' appetitions. Just as perceptions are always perceptions *of* something, so too the monad's appetitions are intentional volitions, desires or appetites for some object or future state, describable only in terms of that to which they refer. Monads' perceptions and appetitions are thus implicitly and inescapably relational, manifesting in actuality the conceptual relations specified in the relational predicates in the monads' complete concepts.

As internal relational qualities, monads' perceptions are the metaphysical ground of the expressive relations between existing monads. Just as the intrinsic or internal relational properties of line L (as longer than M) and line M (as shorter than L) ground the extrinsic or external relation that holds between the two actually existing lines L and M, so too, the internal or intrinsic relational qualities of monads (their perceptions and appetitions) ground the extrinsic or external relations that hold between different, independent actually existing monads. Since it is better that there be something rather than nothing, and indeed more things rather than fewer, provided this does not disrupt the harmonious order, Leibniz believes that God has created as many different monads as was possible consistent with the general order of things.[12] Consequently, the things represented in monads' internal perceptions do correspond to actual, independently existing external things, other monads and their bodies. However, even if God had created only one monad and did not create the others to which its perceptions and appetitions correspond, the internal perceptions of this one existing monad would still be relational qualities, for they would still relate this one monad to all the other possible monads that God could have created. As it is, God did create the other monads. Hence, monads' internal relational perceptions do ground the external relations between existing monads too. As existing or created monads, each is a corporeal substance, a soul or entelechy dominant over the organic body through which it perceives the bodies of others by registering in itself the physical effects of others' bodies on its own. Hence, it is not only as perceiving beings, but also as embodied beings that Leibniz's substances are related to all other substances in the world. All substances are related internally and externally to all the others in Leibniz's thoroughly relational and interconnected universe.

Value revisited

Having established that Leibniz's living substances are inherently relational beings, let us now return to the question of intrinsic value. John O'Neill had identified three senses of 'intrinsic value':

(i)   Intrinsic value as noninstrumental value;
(ii) Intrinsic value founded upon intrinsic (i.e. internal or essential) nonrelational properties; and
(iii) Intrinsic value as real or objective, nonsubjective value.

We have been concentrating our attention on the assumption in (ii) that intrinsic properties are nonrelational properties. I have argued that in the case of Leibniz, the intrinsicality of monads' properties (i.e. the monads' internal essential qualities) does not entail that they are also nonrelational properties. On the contrary, the intrinsic qualities of Leibniz's substances are their internal, relational perceptions and appetitions.

Our question now must be whether the internal *relational* qualities of Leibniz's individual living things are capable of grounding the notion of the individual as intrinsically *valuable*. Do the relational qualities of Leibniz's substances give us reason to regard these substances as having value in themselves? I shall be suggesting that they do. However, the account will require that we reframe questions of value in rather different terms than they have traditionally been posed, for while it is helpful in an ecological context to regard some beings are intrinsically valuable, there is something rather anti-ecological in thinking of intrinsic value in terms of the nonrelational properties of things. Ecology is essentially the biological science of relationships: the investigation of the relations between living organisms and their relations to their environments. Concern about humans' relationships to nature and to nonhuman living beings lie at the heart of environmental and ecological philosophy. Indeed, an ecological philosophy that failed to appreciate the interconnectedness of living things would not be worthy of its name. In light of this, the idea that an ecological theory of intrinsic value should be couched in terms of the *non*relational properties of things begins to look decidedly at odds with itself. Ecophilosophically, intrinsic value is best understood in terms of the quality of an organism's relations to others.

In what follows, therefore, I shall speak of the intrinsic *relational value* of things, as grounded in their internal relational properties. Because many of the relations that one thing has to another are, as it were, instrumentally valuable to the other insofar as they help the other to flourish, we shall have reason to reframe the traditional environmental notion of intrinsic value as noninstrumental value (i). We shall instead arrive at a hybrid notion of the *relational value* of individuals that is at once both (a) *intrinsic* to the individual insofar as it is grounded in its intrinsic or internal relational properties and (b) *instrumental* insofar as an individual's relational value can be assessed in terms of its particular contribution to the overall perfection of the world and to the perfectibility or flourishing of other individuals within that whole.

In chapter 4, we noted a number of features of living and nonliving things that give cause to regard them as things of value. Each living creature is unique and beautiful, each actively seeks to preserve its own life and may be regarded as an 'end in itself'. All are mirrors of the perfect universe without which the universe itself would not be perfect. On the basis of these, we espoused an 'in-principle' bioegalitarianism and as many of these features can be found in inanimate things as well, we had reason also to espouse an 'in-principle' ontoegalitarianism. These features of living and nonliving things are intrinsically valuable in O'Neill's sense (iii): they are real, objective and nonsubjective properties of value.

Relational values, however, are also real, objective and nonsubjective properties insofar as they are grounded in these internal relational properties of things. Some – and perhaps all – of the value-conferring features listed above are relational in character. As a mirror, each creature expresses the perfect whole from its own unique point of view. Beauty is relational both insofar as it is that the contemplation of which gives pleasure to the one who contemplates and insofar as it consists in the ordered harmonious arrangement of the parts of the beautiful thing. The activity or agency of living creatures can be understood only in terms of goals and ends to which the organism strives and to which it stands in relation. Life itself is also fundamentally relational: the very essence or identity of living things, as we have seen, is co-constructed in relation to the essences or identities of others.

The thoroughgoing interrelatedness and interdependence of all things at the basic ontological level of their essential natures suggests that every being has some direct or indirect instrumental value in respect of all other beings. At a deep conceptual level, all things coexist in a state of symbiotic mutual dependence: X depends on Y for the internal relational properties that comprise, at least in part, its essence or individual identity. At the level of existence, Y's existence ensures that X's internal relational qualities or properties – its perceptions – correspond to external reality. We might say, therefore, that both conceptually and existentially, Y has instrumental relational value for X. However, interdependence also stipulates that Y in turn depends upon X for the internal relational properties that are included in Y's essence or nature: X therefore also has instrumental value for Y. Hence, X and Y have instrumental relational value for each other insofar as each is implicated in the formation of the identity and intrinsic relational and value-conferring properties of the other.

This fusion of intrinsic and instrumental value runs counter to the conventional understanding of the terms. The standard view asserts the logical priority of intrinsic value over the instrumental. It is a well-rehearsed point that 'under pain of an infinite regress, not everything can have only instrumental value. There must be some objects that have intrinsic value' (O'Neill 1992, 119). The general claim is that something can have instrumental value only if there is some intrinsically valuable, human or nonhuman, being for which the thing is instrumentally valuable. Humans are instrumentally valuable to dogs only if dogs are intrinsically valuable. The instrumental value of the worms for the bird is parasitic upon the bird's own intrinsic value.[13] However, on the account given above of the formation of the individual identities of X and Y, it is no longer possible to maintain that intrinsic value is prior to instrumental value. Rather, instrumental and intrinsic values, considered as relational values, are inextricably intertwined and mutually supporting.

Instrumental value does require that the thing possesses instrumental value for another being that is itself intrinsically valuable. But equally, and in respect of creatures on both sides of the relation, the instrumentality of the one is responsible in part for the identity of the other, without which the other would have no intrinsic value, nor being, at all. So, not only does the instrumentality of the one require that the other for whom it is an instrument is (or is in process of becoming) a being with its own intrinsic value, but the intrinsic value of each also depends upon its being an instrument for the other. For instance, the worm has instrumental value for the bird, not only as food, but also as an indispensable constituent of the bird's own identity, without which the bird would not be the particular unique bird that it actually is and could not mirror the universe from that particular point or view. Conversely, although the bird has no instrumental value for the worm insofar as the bird regards the worm as a tasty meal, the bird does have a kind of instrumental value for the worm insofar as it, in turn, is an indispensable feature of the worm's own unique individual essence or identity and an, albeit unwelcome, component within the worm's own perceptions and mirroring point of view on the world. The worm would not be the particular worm that it is did it not stand in relation to the particular bird for which it will soon serve as food. The worm's own intrinsic value as the particular living thing that it is, is constituted in part by the bird serving as an instrument in the formation of the worm's identity.

The sense of instrumental value employed here is a wide sense that is perhaps better captured by the notion that every living being and indeed every inanimate aggregate of living beings is 'relationally significant' for all other living and nonliving things. This wide sense of relational value as relational significance applies universally and equally across all living and nonliving things insofar as they are all intrinsically valuable components of the best possible world whose essences are interdependently coformed and coexist with all others in the same world. There is, however, a narrower sense of instrumental relational value that admits of greater or lesser degrees. A being has specific instrumental value for another insofar as it assists or is required in order that the other can achieve its own goals. We have seen an instance of this more specific sense of instrumental value in the example of the worm that serves as instrumentally valuable food for the bird. The bird, on the other hand, lacks specific instrumental value, or has negative instrumental value, for the worm whose own goals are thwarted by the bird's actions.[14] In what follows, we investigate how an individual's instrumental value for others may serve as a marker of its own relational intrinsic value.

Up to now, I have been suggesting that the instrumental value of one being supports, or makes possible, the intrinsic value of *another* being. However, might a being's instrumental value for others also support, or make possible, its *own* intrinsic value? We have seen how the internal, essential or intrinsic properties that constitute the identity or essence of an individual substance or living creature are formed in conjunction with the formation of the identities of others to whom the first is related. Together, they comprise the plurality of beings that is the harmoniously ordered and varied world that God chose to create, having seen that it was the best of all possible worlds. Each individual has an instrumental role to play not only in relation to the world as a whole, but also in relation to each and every individual within it. Could it be then, that the intrinsic value of any one individual is a function of its instrumental value in relation to all the others? Is a being's intrinsic relational value a function, at least in part, of its other-directed instrumentality?

O'Neill's elaboration of 'intrinsic value' in sense (i) – intrinsic value as non-instrumental value – states that something can be said to have intrinsic value when its value is not couched in terms of its having instrumental value in relation to human beings. His citation from Naess reinforces this notion. Framed in this way, the intention is to give weight to the notion that nonhuman living beings have value in and of themselves and to stress that we humans should value them for this reason and not simply on account of any usefulness they have in relation to us. But let us consider the issue of instrumentality from the other side: the usefulness of humans for other beings, human and nonhuman alike. Does not our true human worth reside primarily in our instrumental relational value for other creatures, human and nonhuman alike? Does the relational value of nonhuman creatures also lie in their instrumental value for others? An indication of this line of thought can be found in Leibniz's remark in an early letter to Arnauld. There, he suggests that when faced with choices and questions of relative value, the person who is best equipped to help others is to be preferred over those less able or less willing:

> If the benefits of several people interfere with each other, that person is to be preferred from whose help the greater good in the end follows. Hence in case of conflict, other things being equal, the better man, that is, the one who loves more generally, is to be preferred. For whatever is given him will be multiplied by reflection so as to benefit many people, and therefore many will be helped by helping him.
>
> (To Arnauld, early November 1671: GP I 74; L 150)[15]

I am not advocating that we accept Leibniz's implicit rejection of those who are less well equipped to help others. Nor am I suggesting that we restrict, as Leibniz did, the policy to human beings. After all, every living creature has a positive contribution to make towards the good of others. Instead, I would have us focus on the thought contained in the remark that the ability to help others is a reason to consider that the person has genuine worth. It is this notion that we find expressed by Leibniz when he writes in *On Wisdom* that 'only so much of our life is to be valued as truly living as the good we do in it' (GP VII 87: L 427). As we've seen, every creature has genuine worth insofar as all are valuable constituents of the whole, but the help of others is needed if their full potential is to be realized. It is within our gift to help others as much as we can so that they may best realize this potential. We can choose to promote their flourishing so that their contribution to the whole too may be as good as possible. By assisting others, by using our own instrumentality so that they may flourish and in turn be instrumentally valuable to others, we increase both our own and others' intrinsic relational value. Thus do intrinsic relational value and instrumental relational value come together, such that the greater our instrumental value in relation to the whole, the more we ourselves flourish by realizing our full potential and the greater our own relational intrinsic worth and value as an integrated part of the whole.

At the moment, human beings are not, as a species, faring well on the relational-value scale. Our lifestyles are wreaking havoc and disruption across the globe. We spread ourselves far and wide, pushing nonhuman creatures out as we expand, destroying their natural habitats, refusing to allow them to share our space. If we want to increase our own intrinsic relational value, we should look towards ways in which we might serve as instruments that further the interests of others, human and nonhuman. In short, to increase our own intrinsic relational value, we need to increase our instrumental relational value, learning to live in peace and harmony with fellow humans and nonhuman creatures rather than disregarding their interests in pursuit of what we perceive (erroneously, as we shall see in chapter 6) as being in our own best interests.

However, when we do take seriously our instrumental usefulness to others – which may sometimes mean simply not interfering with the habitats that provide them with food and shelter – we enable them to flourish and to realize their full potential. As a result, their instrumental relational value within the whole is increased as they are freed to play their fullest part in the perfecting of the whole and its constituent parts. For instance, a tree's instrumental value directly correlates with its wellbeing. Its active instrumentality in the unfolding of its relational nature is the expression of its own flourishing nature and intrinsic relational value. The more it flourishes, the better able it is to fulfil its instrumental role within the whole. As carbon sinks and oxygen producers, healthy trees bring immense benefit to others across the whole of the animal kingdom, as well as providing shade, shelter and sustenance for myriad forms of animal and plant life, while humans find peace and relaxation in their company and gain pleasure from the contemplation of their grace and beauty. The internal relational properties of the tree are instrumental properties that underpin its status as an intrinsically valuable part of the whole. Similar remarks would seem to hold for all other living creatures across the Leibnizian world, such as, for example, the birds who benefit from the instrumentality of the tree, but whose seed-distribution activities are instrumentally invaluable to plant life.[16]

As we come to the end of this chapter, let us consider whether inanimate objects might also be granted a degree of relational value that interweaves the notions of intrinsic and instrumental values in a similar fashion. Inanimate objects certainly have instrumental value. The twig has a relational instrumental value for the bird as it builds its nest. The stone has a relational instrumental value for the crow that uses it to crack open the mussel. But do they also have intrinsic relational value tied to this instrumentality? It might be thought inappropriate to extend intrinsic relational value to nonliving things because their instrumentality is not an active instrumentality. However, in chapter 4, we discovered that there is a sense in which even nonliving things may be said to possess an agency grounded in the activity of their constituents. Moreover, the absence of a dominant entelechy and consequent lack of inherent unity did not prevent our regarding nonliving things as objectively beautiful and on that account also intrinsically valuable. The beauty that is manifested in the harmonious order and variety of their parts is a source of pleasure for humans who contemplate it, but beauty is not instrumentally valuable only to humans. It contributes positively to the perfection and goodness of the whole and to other of its parts. Nor does the absence of a dominant entelechy prevent us from considering the nonliving as capable of flourishing, even though it does prevent them from being conceived as 'ends in themselves'.[17] In chapter 6, we shall discover ways of articulating ideas about the flourishing and health of places and spaces and of evaluating places and spaces in terms of the relationships of their constituents. If we consider intrinsic relational value as the ability to flourish in ways that benefit others, the concept of relational value that is both intrinsic and instrumental might well be extended to nonliving materials and environments and thereby allow us to regard them as having an intrinsic relational value that is directly correlated with the objective good that they possess in relation to others. Air of good quality is invaluable not only to humans, but also to the lichen and other living things that depend upon it in order themselves to thrive. Soil that is healthy is able to support the plants that thrive on the nutrients with which it provides them. Healthy ecosystems or communities have relational instrumental value for the living things that inhabit and comprise them. Just as the flourishing and intrinsic relational value of human and nonhuman living things is inextricably linked with their ability to further the goods and interests of others, so too the flourishing or perfecting of inanimate materials and environments is both the source of their relational instrumental value for others and the foundation of their own intrinsic relational value.

In this chapter, I have sketched the broad outlines of a theory of the relational value of living and nonliving things that draws upon the traditional theories of intrinsic and of instrumental value, but which reinterprets intrinsic value as value that is grounded in the internal relational properties that constitute a being's essence and through which it comes to stand in external relation to all other beings in the world. Building on the inherently relational characters of all living and nonliving things, I have attempted to show how the relationally intrinsic and the relationally instrumental are, as it were, two sides of the same coin, each reinforcing and depending upon the other, each individual thing making its own unique contribution to the perfection of the others and to the perfection of the whole of which it is an invaluable part.

Notes

1 Russell 1992, 12–13.
2 Indeed, if, as shall be suggested below, monads' qualities are perceptions and appetitions, all of which are inherently relational qualities, it would appear that the predicates in the complete concept that correspond to these qualities of the created thing are all inherently relational too.
3 Kulstad (1980) argues for the inclusion of relational predicates in the complete concepts of individuals. See also Mugnai 1992, 125; McCullough 1996; and Phemister 2005, 135–138.
4 Theophilus (representing Leibniz) in the header quotation to this chapter makes the same point.
5 Nachtomy agrees with Russell that relational predicates require nonrelational predicates (Nachtomy 2007, 114, 100), but he denies that the former are reducible to the latter. He thus denies the possibility of restating relational predicates using nonrelational predicates or conceiving all the relational qualities of actually existing substances in terms of nonrelational qualities.
6 See chapter 2, p. 42.
7 *Principles of Nature and of Grace* §2: GP VI 598; AG 207.
8 Bradley (1920 [1893], 25–34) uses this fact as part of his critique against pluralist metaphysics. A quality, he argues, can stand in relation to another quality only if it can be identified independently of the relation, which cannot be since it can only

exist through the relation. Relations, he maintains, are therefore contradictory: it is impossible for there to be relations without terms that stand at either end of the relation, but conversely, there are no terms unless we identify them by distinguishing them from other terms to which they therefore must stand in relation. Hence, there can be no relations without terms, but equally, there can be no terms without relations. This double bind is the reason why Bradley assigns all relations and their supposedly independent, nonrelational terms to the realm of mere appearance and asserts that relations and relational ways of thinking cannot characterize the true nature of reality. I discuss a Leibnizian response to Bradley's dilemma in Phemister (2016).

9 *Principles of Nature and of Grace* §2: GP VI 598; AG 207.
10 They also stand in conceptual relation to essences in nonexisting possible worlds: 'our soul expresses God, the universe, and all essences, as well as all existences' (*Discourse on Metaphysics* §26: GP IV 451; AG 58).
11 *Discourse on Metaphysics* §9: GP IV 434; AG 42.
12 *On the Ultimate Origination of Things*: GP VII 303–304; AG 150–151. See also Phemister 2005, 149–152.
13 The basic intuition here is the one that underpinned Russell's prioritizing of nonrelational monadic predicates over relational nonmonadic predicates: the belief that one must be able independently to identify the individuals themselves prior to considering them in relation to each other. However, just as matters proved not to be quite so simple in the case of individuals and their complete concepts, so I would suggest, the priority granted to intrinsic value over instrumental value is not all that it at first seems.
14 In the wider scheme of things, of course, the bird's actions precipitate the next stage in the unfolding of the worm's essence and of the essences of the substances that currently comprise the worm's body. The worm and the substances that make up its body are ready to unfold, transformed as parts of the bird's body (chapter 4, pp. 70–72).
15 See also chapter 4, p. 80; chapter 6, note 40 (p. 129).
16 Of course, creatures' flourishing is not an uninterrupted upward journey. The worm's flourishing plummets dramatically when it serves as food for the bird. Nevertheless, to the extent that it was healthy before it was eaten, the greater was its instrumental value for the bird.
17 In chapter 7, we discuss Leibniz's views on the perfections of bodies. Our focus there is living bodies, but many of the points raised can also be used to support a claim to the perfecting of inanimate things that makes room for the development of a theory of ways in which they too might be said to flourish.

6   Space, place and value

Sometimes I think that these spaces we inhabit are not physical places at all, just layer upon layer of memories. They are built out of experience – human experience – not steel or pre-cast concrete.

(Jonathan Coe, BBC Radio 4, 7 October 2011)

In chapter 5, we discovered that the individual substances that comprise the Leibnizian universe are deeply relational beings. The complete concepts that encapsulate their unique identities and essences specify the internal relational properties that are the basis of the external relations that the creature will have to others when all are created to form the actual world. The identity of any one possible being is shaped in conjunction with the identities of all the possible beings whose perceptions, when each entity is created, reflect the world they simultaneously jointly comprise and share. Each and every living creature is 'world-bound' insofar as a complete articulation of its essence would include reference to all other things in the same world. It is not possible to conceive of an individual without conceiving it as part of the (possible or actual) world in which it occurs. Each individual reflects or mirrors the world to which it belongs. In this respect, perceiving souls, are essentially representative beings. Every living organism has a soul or entelechy whose perceptions reflect the world to which it belongs. Did its perceptions reflect another possible world, in which the environment it perceives itself to inhabit is different, then the individual too would be a different individual with a different nature or essence. Particular places or environments are not mere backdrop locations we happen to pass through, as if we could be transported from one place to another without losing our core identities. The environments we inhabit – both our immediate environment and the wider environment that is the world itself – are the environments reflected in our perceptions, the internal relational qualities that are integral to our very identities.[1] As Naess also recognized,[2] we change our environments, but equally, our environments and their inhabitants change us.[3] Everything that happens in our environments is formative of our particular identities and similarly our actions affect everything around us, rippling through the whole of creation.[4]

Environments are the particular places we inhabit. In this chapter, we consider Leibniz's understanding of space and the particular places that comprise it.

Taking cognizance of his view that all living souls or entelechies are embodied,[5] we will be considering space and place on two levels. First, we consider Leibniz's theory of physical space and place as arising from the relational qualities of coexisting organic and inorganic extended bodies. Later in the chapter, we ask whether we can also conceive of space and particular places within space in terms of, and as grounded in, the relational appetitive and perceptive qualities possessed by souls or entelechies. While this goes beyond what Leibniz himself proposed, it follows naturally from his ontology of interconnected beings and his relational account of physical space. It has the advantage of opening up ways of conceiving space and places not merely as measurable quantities of shape, size and distance, but also as qualitatively rich, imbued with political, ethical, aesthetic and spiritual value.[6]

Kingdoms of nature

Being both embodied and ensouled, corporeal substances' actions are explicable both in terms of the mechanical movements and resistances of their bodies and in terms of the perceptions and appetitions of their souls or entelechies. In more technical terms, every event is explicable by both final and efficient causation, each system of explanation being complete in itself, making no use of the conceptual framework employed by the other. Thus, the living creatures in the Leibnizian world belong at one and the same time to what he calls the natural kingdom of efficient causes and to the natural kingdom of final causes. The first comprises bodies, acting and being acted upon in accordance with the laws of motion and resistance. The second comprises all perceiving, appetitive souls acting in accord with the laws of final causes that lead them invariably to pursue what they perceive as good or beneficial and to shun what they perceive as bad or harmful (*Monadology* §79: GP VI 620; AG 223).

Focusing on himself as an ensouled *body*, Leibniz explains that his present act of writing is explicable physically as the result of mechanical interactions between bodies. Were this explanation complete, it would take in the prior history of the entire physical world, although more immediate causes would include the current operation of his own body, including the heart pumping blood through his body and, as we would today conceive, the neurons firing in his brain. Equally, however, focusing on himself as an embodied *soul*, his writing is conceived as the result of all the appetitions and perceptions that have led from the moment of his creation up to the present moment. Thus,

> [t]here is an infinity of past and present shapes and motions that enter into the efficient cause of my present writing, and there is an infinity of small inclinations and dispositions of my soul, present and past, that enter into its final cause.
>
> (*Monadology* §36: GP VI 613; AG 217)

The example is generalizable to all living creatures. Every soul or entelechy has desires and appetites, aims at certain goals and acts always in pursuit of what, often subconsciously, it perceives as conducive to the preservation of its being. The parasite searches for a hospitable host; the dog seeks a warm, dry comfortable place to sleep. Meanwhile, creatures' bodies manifest these desires as the motion of their bodies. Hence, the parasite's body and the dog's body each move in ways that will, it is hoped, result in the satisfaction of their respective desires. As explained in chapter 2, each monad's primitive active force is modified simultaneously both psychically and physically, so that each creature acts both in accordance with its soul's or entelechy's goals and aspirations (by final causation) and by the movement of its organic body (by efficient causation).[7]

In contrast with the activities of the parasite and the dog, Leibniz's present writing is an example of free and self-consciously chosen action. In theory, Leibniz is aware of his decision to put pen to paper and could have chosen not to do so. In this respect, Leibniz assumes responsibility for the effects of his actions. It marks him out as a rational, self-consciously aware, free moral agent and grants him citizenship of a further kingdom identified by Leibniz as the moral kingdom of grace. Only free and rational minds, able to recognize the final causes by which they act as goals or ends they have themselves freely chosen to pursue, belong to the moral kingdom of grace.[8]

> Since earlier we established a perfect harmony between two natural kingdoms, one of efficient causes, the other of final causes, we ought to note here yet another harmony between the physical kingdom of nature and the moral kingdom of grace, that is, between God considered as the architect of the mechanism of the universe, and God considered as the monarch of the divine city of minds.
>
> (*Monadology* §87: GP VI 622; AG 224)

Leibniz here distinguishes sharply the two kingdoms of nature (efficient and final) and the moral kingdom of grace, but it is clear that all three kingdoms are in fact 'natural' in a wider sense of the term. Although he does not explicitly describe the kingdom of grace in this way, it must be situated within the natural world, for this moral kingdom, comprising minds self-consciously following their own final causes, is obviously a subset of the natural kingdom of final causes – a kingdom that encompasses *all* beings that act by final causes, whether self-consciously or not. 'Nature' may be regarded as the conjunction of the kingdoms of efficient and of final causes. Within this nature, a further kingdom of grace is then set against the kingdom of efficient causes (the 'physical kingdom of nature'). As a subset of the natural kingdom of final causes, the kingdom of grace itself is evidently part of a greater 'nature' that encompasses all three kingdoms, efficient, final and grace. In this sense, the moral realm of grace is not above and beyond the natural world, but belongs within it. Taken as a whole, the natural world is *both* an extended physical world of moving bodies and their efficient causes *and* a world of perceiving, desiring and willing beings and their final causes, some of which are recognized as personal goals by minds aware of themselves and capable of acting responsibly and freely.

It is in respect of the natural world as a physical world of bodies that Leibniz developed his relational theory of space, according to which space and individual places emerge from the relations between bodies, relations that are themselves founded upon bodies' relational qualities.[9] The theory is outlined in the following section.[10]

Extended relational space and places

In his correspondence with Samuel Clarke, Leibniz defended a relational theory of physical space, providing an alternative to the Newtonian view that space is absolute and could therefore in principle be devoid of any objects whatsoever. Leibniz regards absolute space as a mere fiction or abstraction and proposes instead that space is 'nothing else but an order of the existence of things, observed as existing together' (Leibniz's Fifth Letter to Clarke: GP VII 395; LC 63). Space, considered only in itself, is 'an ideal thing' (GP VII 396; LC 64). As such, it describes an abstract relationality, or, as we may put it, an abstract geometrical structure, that becomes real only on the creation of things whose relational qualities underpin the external relation that exists between the relata. Leibniz explained this to Clarke in the example of the ratio of the two lines L and M that we examined in chapter 5.[11] Lines L and M possess respectively the relational qualities of being longer and being shorter than the other, but it is only when the lines are real, that is when they actually coexist, that the relation between them is realized and space itself, as the relation between them, comes into being as something concrete and real (GP VII 395; LC 63). Actual space requires particular existing things or substances whose relational qualities establish the order of coexistence that in turn constitutes space.

However, for Leibniz, as we have seen, the natural world is in flux. Nothing remains the same from one moment to the next. Relations between bodies are constantly changing. Does this mean that space and individual places are also constantly changing? Space itself is simply that 'which results from places taken together' (Leibniz's Fifth Letter to Clarke: GP VII 400; LC 70). Speaking strictly, however, no place remains the same even from one moment to the next. Yet, we think of space and places as constants. We think of ourselves as moving through space from one place to another and believe that it is possible to return to the same places as we were before, as when we speak of returning to the places of our birth or to old haunts. Fortunately, despite flux, it is possible to accommodate our familiar everyday conceptions of space to some extent. That some changes of relations between the constituents of a place are slower than others is enough, Leibniz thinks, to support at least a feigned supposition that, 'among those co-existents, there is a sufficient number of them, which have undergone no change'. We can then go on to stipulate that 'those which have such a relation to those fixed existents, as others had to them before, have now the *same place* which those others had' (GP VII 400; LC 69). Thus, we may be said to return to the same place, as when we re-enter a room, because the changes in the relations between the walls of the room remain unchanged relative to other rooms nearby.

On Leibniz's relational theory, the subjects whose relations give rise to real space and its component places are physical objects, living and nonliving bodies whose relational properties include such things as their shape, size, location and locomotion relative to other bodies.[12] This space is replete with material objects and organic bodies whose mechanistic activity is explicable by efficient causation, governed by the physical laws of motion and collision. Physical space is described in quantitative terms. Mathematics and physics describe its topology. The bodies whose relations form the core of relational space are conceived as physical objects extended in length, breadth and depth. Even in respect of the relations between living organisms – the 'living machines' that are the organic bodies of corporeal substances – it is the physical, geometric relations between them that are taken as constituting space. No reference is made to the perceptual, affective and conative relations between the corporeal substances' souls or entelechies.

It is entirely possible to proceed in the environmental sciences through study of the purely physical interconnections among biological entities, to engage in ecology but not also in ecophilosophy.[13] It is perfectly understandable that this be the case. Not knowing from the inside, so to speak, how other beings experience the world, we fall back to concern for their physical wellbeing and invariably resort to purely scientific management of them and of the places they inhabit. All the same, this comes at a cost, especially if taken so far as to allow us to forget or even deny that nonhuman living things are capable of experiencing and feeling the world from their own distinctive and unique perspectives. It is in this regard that the importance of combining ecological science with the more speculative insights of the ecophilosophers can be seen. Ecophilosophy allows us go beyond the confines of experimental, natural-scientific quantitative methodology and to reintroduce teleological, ethical and aesthetic relations to our assessment of the condition and wellbeing or flourishing of things that comprise the natural world. In chapter 7, we explore ways in which the physical states of creatures' bodies serve as guides to their psychical wellbeing. In this chapter, we consider the environments of these creatures as psychical places that arise from the perceptual, affective and conative relations between living beings. The resulting account of relational psychical space and places lends metaphysical support to conceptions of space and place, more familiar in the humanities than the natural sciences, as places that embody values, spiritual, ethical and aesthetic.

Leibniz's philosophy is particularly amenable to adaptation along these lines. He conceives the natural world in both physical and psychical terms. On his view, each living thing is both embodied and ensouled. The world is both a natural realm of bodies and efficient causes and a natural realm of souls and final causes. By focusing on the world as a realm of final causes, it ought to be possible to supplement Leibniz's theory of physical or geometric space with a theory of psychical or value-laden space. Where the former is grounded in the quantitative relations of bodies, the latter is grounded in the qualitative perceptual and conative qualities of minds, souls and entelechies. Just as it is possible to explain the world either as a system of efficient causes or as a system of final causes, it ought to be possible to describe space either as an ordering of coexisting bodies or as an ordering of coexisting souls and entelechies. And just as there is harmony between the realm of efficient causes and the realm of final causes, so too, we can expect to discover a harmony between space as the relations of bodies and space as the relations of souls.

The Leibnizian account of relational physical space based on the *physical* characteristics of living corporeal substances will serve as our model for the construction of a parallel 'value-space' grounded in the relationality of the bare feelings, sensations, perceptions, appetitions, impulsions, desires and volitions that characterize the *psychical* aspects of living beings. But before exploring this in more depth, it will be helpful to recall and to expand briefly upon the psychical dimensions of Leibniz's universe.

Perceptions and appetitions

As we discovered in chapter 5, the internal mental qualities belonging to Leibnizian monads or simple substances are their perceptions (sensations, imaginations, thoughts as well as feelings of pleasure and pain, emotions and passions) and their appetitions (impulses, appetites, desires, volitions).[14] All are inherently relational. Our perceptions are always perceptions of something; our appetitions are always focused on a desired object or possible state of affairs. Through its perceptions, each mind, soul and entelechy represents everything in the universe. Perceiving minds, souls and entelechies are like mirrors that represent the whole universe. Hidden deep in our representative states are perceptions even of what is happening on Jupiter and Saturn, the ocean beds and between the minute particles of the earth.[15] Nothing happens in the universe without it being registered at some level in our mental state.[16] Correspondingly, everything that is happening to us is registered in turn in the psychical states of all the others. I perceive the gull as it flies past, but so too, it perceives me. As I walk, I subconsciously register the tiny organisms within each grain of sand, but they too insensibly perceive me stepping on them. In the words of David Abram in a discussion of Merleau-Ponty,

> To touch the coarse skin of a tree is thus, at the same time, to experience one's own tactility, to feel oneself touched *by* the tree. And to see the world is also, at the same time, to experience oneself as visible, to feel oneself *seen*.
> (Abram 1996, 68)

Mutual representation of all by all is important, but so too is the manner in which we perceive the world. For the most part, we perceive things clearly, but confusedly. We perceive as wholly inanimate, objects that are, in Leibniz's opinion, aggregates of living things, for we do not clearly discern their component parts. When we perceive a thing both clearly and distinctly, we are able not only to recognize the thing as an individual thing, but also recognize in it the features that distinguish it from others. When we perceive another clearly and distinctly, we clearly perceive the distinguishing marks that individuate it from ourselves and from others.[17] So, for instance, I perceive the gull clearly and distinctly when I recognize not only that I am perceiving it from my own distinctive point of view, but also that it is perceiving me from its own equally distinctive point of view. The more that I perceive another clearly and distinctly, the more I recognize its distinctive otherness and appreciate that it perceives the world in its own way and that it has its own distinctive desires and needs.

Whether perceptions are distinct or confused is a function of the appetitive force responsible for the transition from one perception to the next. How clearly and distinctly I perceive the universe depends upon the amount of appetitive force propelling my mind onwards. The greater the force, the more distinctly the universe will be represented in the ensuing perception.[18] Particularly pertinent to our discussion here is the further observation that appetition is also, by an indirect route, responsible for the attitude that we take towards that which is represented in our perception, as for instance, whether we regard the things we perceive with compassion, care and love or with disdain, indifference or even hatred. Why this should be so lies in a set of connections Leibniz discerns between the perception of beauty, order and perfection and feelings of happiness, joy and love.

Leibniz locates the source of human joy in the distinct perception of nature's diversity, order, interconnectedness and sheer magnificence. He emphasizes the fact that when we learn to appreciate the order and harmony of the world, it pleases us. We experience joy when we are aware of this harmonious order. '*Joy*', he writes, 'is a pleasure which the soul feels in itself' (*On Wisdom*: GP VII 86; L 425). As a pleasure, joy is a 'sense of perfection'; pain, on the other hand, is a 'sense of imperfection' (*New Essays*: A VI vi 194; RB 194).[19] Happiness, meanwhile, is a 'state of permanent joy' (*On Wisdom*: GP VII 86; L 425).[20] By this, Leibniz does not mean that the happy person is ecstatically happy all the time. The happy person does not necessarily enjoy a permanent state of *occurrent* happiness. Rather, happiness is found in the movement towards new pleasures.[21] A person who is happy is one whose overall disposition tends more towards future pleasures than future pains.

Single pleasures are merely steps on the way to lasting happiness (*New Essays*: A VI vi 194; RB 194). Each single pleasure is a

> feeling of a perfection or an excellence, whether in ourselves or in something else. For the perfection of other beings also is agreeable, such as understanding, courage, and especially beauty in another human being, or in an animal or even in a lifeless creation, a painting or a work of craftsmanship, as well. For the image of such perfection in others, impressed on us, causes some of this perfection to be implanted and aroused within ourselves.
>
> (*On Wisdom*: GP VII 86; L 425)

We take pleasure from our perception of the beauty and order of the universe and are pained when we perceive its imperfection, disorder or destruction.[22]

Beautiful things themselves, as we saw in chapter 4, Leibniz defines simply as those the 'contemplation of which is pleasant' (*Elements of Natural Law*: A VI i 464; L 137).[23] However, although we are pleased by beautiful things, Leibniz holds that it is only when beautiful things are themselves capable of happiness that we are able to love them, for 'to love is to be disposed to take pleasure in the perfection, well-being or happiness of the object of one's love' (*New Essays*: A VI vi 163; RB 163).[24] Love, he goes on to explain,

> involves not thinking about or asking for any pleasure of one's own except what one can get from the happiness or pleasure of the loved one. On this account, whatever is incapable of pleasure or of happiness is not strictly an object of love; our enjoyment of things of that nature is not love of them, unless by a kind of personifying, as though we fancied that they could themselves enjoy their perfection. When one says that one loves a fine painting because of the pleasure one gets from taking in its perfections, that is not strictly love.
>
> (*New Essays*: A VI vi 163; RB 163)[25]

Defining love in this way entails that feelings of love may be directed only towards rational beings, for only they are capable of happiness and the joyful experiences that arise from the contemplation of beauty. However, many will consider this definition far too restrictive. After all, do we not love at least some nonhuman animals, especially those that are our pets? Even Leibniz acknowledges that some people seem to love animals insofar as they seek a creature's good in itself and do so not, or not only, for human gain. Leibniz tries to accommodate his theory to these types of loving experiences by suggesting that such people are able to feel love for nonhuman animals only because they consider animals as rational to some extent: animals' sensations qualify them as having some 'reasonable element' (*Elements of Natural Law*: A VI i 465; L 137).[26] However, *pace* Leibniz, rationality in animals need not be the ground of our love for them. For might we not also feel love towards them because we recognize that they too have feelings and sensations and are capable of pleasure and pain?[27] Even though they may be incapable of experiencing the kind of happiness that arises from the perception and contemplation of beauty and order, we nevertheless derive pleasure from seeing them thrive and from seeing their needs met.

Might it not also be possible to have loving feelings towards living things whose experiences do not amount to full sensations of pleasure and pain, towards those whose perceptions are only bare dreamlike or faint feelings? Although they do not experience pleasure or pain in the fullest sense, they are capable of responding when dramatic changes occur in their environments, which in turn suggests they have something akin to desires or what Leibniz calls 'semi-appetitions' to pursue or to flee. Like the larger animals, they too have needs and interests that serve their wellbeing. From bees to bacteria and beyond, all are capable of flourishing or perishing. If one seeks, and finds disinterested pleasure, in their wellbeing might not one's feelings be described as a kind of love?

Instead of denying feelings of love towards nonhuman beings, we can admit degrees of intensity in our loving feelings towards the nonhuman, or even towards some humans over others. We do not love all to the same extent. We mourn deeply for our pets, but we do not grieve so much at the death of, say, a fly. Our most intense feelings of love – what Leibniz considers as the only kind of love – are usually directed towards our fellow human beings, to those similar to ourselves in their capacity to attain true happiness. Happiness signals the perfecting of the human. It is our perception or contemplation of other humans' happiness that arouses in us the pleasurable feelings that cause us to love them. But nonhuman animals and other living creatures are not without perfections too. Indeed, in chapter 4, we found their active life force, perceptual sensitivity (to varying degrees) and beauty (order and variety) supported 'in-principle' bioegalitarianism and I highlighted Leibniz's assertion that 'every substance bears in some way the character of God's infinite wisdom and omnipotence and imitates him as much as it is capable' (*Discourse on Metaphysics* §9: GP IV 434; AG 42). Just as 'love' describes the feelings aroused by the perception of human perfections, the same term can, with due recognition of varying degrees of intensity, be used to describe the feelings of care and compassion that arise from the pleasurable feelings we get when contemplating the perfections of nonhuman beings.

Inanimate things too have perfections. In chapter 4, we were able to attribute value to the nonliving as well as to the living on the basis of their beauty and other perfections.[28] The person who claims to love a fine wine or a beautiful painting is signalling that these things deserve to be treated with care and consideration as things of value, not because they please us (although they do), but because they are intrinsically valuable in themselves,[29] contributors to the net perfection of the universe. Again, while not the highest form of love, and no doubt less intensely felt, the sorrow one feels on breaking an item of great sentimental value or an object of great beauty and the care that one takes to preserve them from damage suggest feelings towards them not unlike those of love. That even the inanimate should not, as far as possible, be harmed or treated with disdain or disrespect is an attitude the cultivation of which could in theory transform our 'throwaway' society into one that treats all things, living and nonliving, with care and respect. In what follows, I shall concentrate on love for animate beings, but with the proviso that, albeit to a lesser degree, love can be extended also to the inanimate.

In order to resist the notion that only rational beings (e.g. humans, angels, God) can be proper objects of love, I propose a modification to Leibniz's definition of love, one that does not require attributing a capacity to reason to animals or to other living things. Instead of defining love as the taking of pleasure in the 'perfection, well-being or happiness' of the other,[30] we can define love as the taking of pleasure in the 'perfection, well-being or specific good of the other'.[31] This revision allows for the possibility of feelings of love towards animals irrespective of whether we believe them to be rational and thereby capable of happiness. Fish, plants and indeed all organisms can also be described in terms of their perfection, wellbeing and what promotes their individual good. Leibniz's pluralist panpsychism postulates myriad living beings throughout the physical world.[32] Bacteria, fungi and other microorganisms, atoms and subatomic particles, we presume, do not feel pain and pleasure to the degree experienced by self-conscious beings. However, we have seen that all are mirrors of the universe and that as perceiving beings, they mimic God's omniscience to some extent. Their bare feelings bear some resemblance to conscious affective states, leading them to move in certain directions as they pursue the pleasant and avoid the unpleasant. And so, while we may never feel for these organisms the kind of intense love we feel towards our fellow humans and larger animals, our modified definition of love as the taking of pleasure in the 'perfection, well-being or specific good' of the other, allows for the possibility of an unselfish caring love directed towards *all* living creatures, benevolent love that is shown in the desire to promote their wellbeing, motivated simply by the disinterested pleasure we gain from seeing other living creatures thrive, not by anthropocentric and utilitarian concerns.[33]

Justice

Love is the handmaiden of justice. 'Justice', writes Leibniz, 'demands that we seek the good of others in itself, and since to seek the good of others in itself is to love them, it follows that love is of the nature of justice' (*Elements of Natural Law*: A VI i 465; L 137).[34] In the preface to his *Codex Juris Gentium Diplomaticus*,[35] Leibniz distinguishes various degrees of justice or 'natural right'. First, commutative justice or '*strict right*' stipulates that, as far as possible, no one is to be harmed. Its aim is the preservation of peace, but it does not demand that we positively promote the good of others. Second, distributive justice or '*equity*' promotes positive moral obligations to treat everyone fairly, not in an abstract sense of equality, but rather in the sense of giving everyone what he or she deserves: 'it commands us to *give each one his due*' (GP III 388; L 422). These two degrees of justice, are, as it were, secular. The third, and highest, degree of justice – universal justice or '*piety*' – operates under God's laws. Universal justice goes beyond human law. Human law, for instance, does not prohibit us from abusing our own bodies or damaging our own property so long as this harms no one else. However, these things are prohibited by God's universal laws:

> duties that do not otherwise seem to concern others, as for example, not to abuse our own bodies or property, though they lie beyond human laws, are yet prohibited by the law [*jus*] of nature, that is, by the eternal laws of the divine monarchy, since we owe ourselves and our all to God. For if it is of interest to the state that no one should make bad use of his property, how much more to the interest of the universe!
>
> (Preface to the *Codex*: GP III 389; L 423)

Universal justice commands that we live honourably, virtuously or piously, acting always for the good of all (GP III 389: L 423). It requires that we aim to acquire the 'custom of acting in conformity with reason which makes virtue a pleasure and second nature to us' (*New Essays*: A VI vi 188; RB 188).

Leibniz understands distributive justice narrowly as 'charity' (Preface to the *Codex*: GP III 387; L 422), but universal justice is the 'charity of the wise' or 'universal benevolence' (GP III 386; L 421).[36] '[C]haritable people love their neighbours with some measure of tenderness, they are sensitive to the good or harm of others' (*New Essays*: A VI vi 215; RB 215). In the wise, this love becomes universal, which for Leibniz means that it is extended to all rational agents. The wise seek universal harmony and that others achieve perfection through attainment of wisdom.[37] Although no finite being is perfectly wise, the notion is useful as an ideal towards which to strive (Hostler 1975, 53).

The wise person loves God above all,[38] finding the highest pleasure in the contemplation of the perfect being:

> Since God is the most perfect and happiest, and consequently, the substance most worthy of love, and since *genuinely pure love* consists in the state that allows one to take pleasure in the perfections and felicity of the beloved, this love must give us the greatest pleasure of which we are capable whenever God is its object.
>
> (*Principles of Nature and of Grace* §16: GP VI 605; AG 212)

> All *pleasure* is a feeling of some perfection; one *loves* an object in proportion as one feels its perfections; nothing surpasses the divine perfections. Whence it follows that charity and love of God give the greatest pleasure that can be conceived.
>
> (*Theodicy* §278: GP VI 282; H 297)[39]

Loving God, the wise also seek to bring about God's will – the perfection of the whole – for to love God, 'the seat of universal harmony', is to love that which God has created. The love of God and the love of others are one and the same: 'it is the same to love truly or to be wise, and to love God above all things; this is to love all or to be just' (to Arnauld, early November 1671: GP I 73; L 150).[40] Indeed, since those who love take pleasure in promoting the happiness and perfection of others, and since human perfection resides in wisdom and justice, the true good of the wise and just is inseparable from the true good of those whom the wise love.[41] In keeping with Leibniz's definition of love, at the highest level of justice love must be extended to all of humanity, or to all rational beings. However, by our revised definition of love proposed above, at the highest level of justice love must also be extended beyond humanity to every living thing in creation and to the truly universal harmony of all with all. Accordingly, we may amend Leibniz's remark to Arnauld (GP I 73; L 150) to read: 'it is the same to love truly or to be wise, and to love God above all things; this is to love all *living things* or to be just'.

Value relational space

We are now in a position to return to the topic of space and the consideration of space as a relational structure of coexisting souls or entelechies. Space considered on these terms results from relations of souls and entelechies founded upon their respective internal relational qualities, namely, their perceptions and appetitions. Mechanistic efficient causation governs the ways that bodies act and react to each other and is characteristic of the relations between bodies, but it is teleological final causation that is operational in the case of souls and entelechies as they seek what, whether obscurely, confusedly or distinctly, they perceive as good. The dimensions of space and places as physical are measurable quantities, but the dimensions of space and places as psychical are neither measurable nor quantitative, but rather qualitative. Psychical space is a space where values such as goodness, beauty, perfection and justice are paramount.

It might be thought too radical to think of space and places in such terms. Many will prefer to regard the psychical dimensions of space as purely metaphorical. However, the living beings whose bodies are related in such a way as to comprise physical space are not metaphorical beings. They are real, embodied, perceiving, appetitive beings. The relations between their perceptions and appetitions are as real as the relations between their organic bodies. To think of space and places as arising from the relations between these psychophysical beings (rather than just as arising from the relations between their bodies) is to consider space and places as both quantitative and qualitative and to regard their psychical dimensions as more than mere metaphors.

Leibniz holds that actual physical space comes into being when bodies stand in relation to each other. The same may be said in the case of a relational space of value. It too exists only provided there are perceiving, appetitive beings standing in relation to one another. Just as there is no extended empty space before there are physical bodies, so too, in its psychical dimensions, space contains no just, compassionate or loving places until there are actual perceiving beings who act justly and with compassion and love. It contains no places of safety until there are beings relating to each other in ways that are supportive and health-promoting, nor places of terror unless there are belligerent beings hell-bent on each others' destruction. In themselves, love and justice are abstract concepts, ideal and unreal. What are real are the living beings whose relational qualities determine whether the places they inhabit are grounded in relations based in love, kindness and openness or in hatred, envy and secrecy.

The overall character of space as a space of values is formed from the nature of the perceptual and appetitive relational qualities of the coexisting entities that comprise it. Perceptions are bare feelings, sensations or cognitive states in which the world is represented, but perceptions are also affective states, pleasurable or painful, happy or miserable to varying degrees. Appetitions are the modifications of the monad's primitive force by which the soul or entelechy progresses from one perceptual state to the next.[42] Working together, the soul's appetitions direct the soul towards what it perceives, rightly or wrongly, as good and therefore likely to provide it with pleasurable future perceptions. No creature is ever entirely ambivalent to what it perceives. Living beings always prefer to change their perceptions in one way rather than another. For instance, the dog prefers to get a closer look at the cat rather than the stick; the flea prefers to feel itself buried in the hair of the dog rather than on the leaf of the plant. Each living thing seeks its own ends and, in line with the laws of final causes, pursues those things and situations that appear to it as being in its own best interests. In the vast majority of cases, the creature is not consciously aware of its goals. The worm has appetitions towards certain things rather than others. If it feels hungry and tired, it will pursue food and rest, but it does not do so consciously; it is not conscious of these goals as its own. The worm merely seeks, adopting Spinoza's terminology, to preserve its being through its own *conatus*.

However, just as in extended space, the resistance or impenetrability of bodies entails that a body already in a particular place must be moved before another can enter the same place, so too, the desires and appetites of living things often conflict with one another, making impossible the simultaneous satisfaction of both. The more powerful wins to the detriment of the less powerful. If the eagle is to be fed, the mouse is sacrificed; but the mouse desires to preserve its life and if it succeeds, the eagle goes hungry. If we were to apply Leibniz's hierarchy of justice to cases like these, we would say that the relation of the eagle and the mouse conforms to the level of 'strict justice' according to which *as far as possible*, no harm is to be done to others. The eagle needs to eat and would not harm the mouse were it not hungry. Although exceptions can always be found,[43] for the most part, the animal kingdom operates in accordance with this rule. As Freya Mathews points out in a slightly different context, '[a]nimals do not follow the so-called law of "dog eat dog"' (Mathews 1991, 157). Humans, however, often take this rule too far, regarding as basic needs that are not actually so. We neglect to temper our self-interestedness through the application of the next level of justice, justice as equity. This level of justice requires us to treat people and other life forms with fairness and to recognize that their right to life may overrule our human desires for material wealth, leisure and comfort. At the highest level of justice, the wise follow the principle of universal benevolence, although even the wise cannot always bring good to all. Nonetheless, the wise man who loves all 'necessarily strives to please all, even when he cannot do so, much as a stone strives to fall even when it is suspended' (to Arnauld, early November 1671: GP I 73; L 150).[44]

Competing and noncompeting desires (akin to competing motions and resistances of bodies) may be conceived as adding a qualitative dimension to quantitative space, a value dimension that is founded upon the perceptual and appetitive relational qualities belonging to the living things that make up the Leibnizian universe. All individuals acting, consciously or unconsciously, in accordance with final causation together comprise a universal value space, the relations between them determining the particular character of this space as they react positively or negatively to each other according to whether the other is perceived as friend or foe, as desirable or undesirable. However, only rational minds can freely choose to conduct their relations to other rational and nonrational beings either in a spirit of universal benevolent love, justice and charity or in terms of hatred, fear and competitive self-interest.

Competition is common not only between two individuals, but also between and within groups of individuals, such as nation states and communities, including ecocommunities. Modelled upon extended *places* within extended relational space, such groupings and communities may be understood as *places* of value within the universal space of values. It is to these that we now turn.

Value relational places

Just as physical space is composed of smaller physical places, so too psychical space comprises smaller psychical places. Because every organic body is attached to a dominating mind, soul or entelechy with which it makes up a corporeal substance, every extended physical place is also a psychical place, a place of values, describable in terms of the affective feelings, sensations, cognitions, desires, interests, attitudes and needs of related living beings. The beach is not just a place of material objects; it is also a place of life – a place where the birds perceive, the constitutive parts of each grain of sand have some bare feeling, and where the flora and fauna perish or thrive according to the nature of their relations to the other beings. All communities of living individuals constitute places of value, whether this be a rainforest, filled with an incredibly diverse array of life forms from termites, insects, birds and primates to all manner of trees and plants, or a city, full of different, but equally diverse, conjunctions of species. The particular goals and desires of the individual living things within each community need not always coincide with those of the others – every individual pursues its own goals – but nonetheless, each community has an identifiable structure and order and at some level, the majority of individuals within it desire, albeit for the greater part not consciously, the continuation or survival of the whole. The structure and order is what allows us to talk of such communities as ecosystems or ecocommunities,[45] that is, as habitats within which individuals cooperate and compete with one another as they each seek to further their own individual goals and outside of which these goals could not be realized.[46] Indeed, some members may discover that their goals cannot be realized in a particular environment, perhaps because of human pollution, changing environmental factors or simply because they have strayed from their natural habitat. Some creatures are able to adapt to new conditions, rather as a square peg may be hewn in order to fit a round hole. Others, such as birds that migrate in winter to warmer climes, leave in search of better conditions. Others again, victims of war or famine, do not survive in their present form, their perceptions and appetitions entering into the period of stupor that for Leibniz characterizes death.[47]

Despite this flux as individuals arrive and depart, the places themselves, just like their physical counterparts, may be said to remain relatively stable. So long as there remains sufficient stability, grounded in the common goals and ends of most of the beings whose relational qualities act as the foundation of the place, we may suppose that the value place remains the same. A particular institution, such as a hospital, would not exist as anything more than an extended place made up of buildings, wards and theatres did not the medical practitioners, administrators and patients work together in pursuit of healing. The goals and aims may remain relatively constant and allow us to construct the notion of the hospital that is more than its mere buildings, even though individual members of staff and patients come and go over time. Just as the place in the park is considered to be the same place despite the removal of a bench, so too, a place constituted by values and goals, such as the hospital, remains the same on account of the goals and aspirations of its fluctuating membership.

All the same, although every physical place is also a place of value, the converse does not hold: not every place of value corresponds to a determinate physical place. Common goals, shared cultural, ethical and spiritual values can form communities and institutions that are not confined to particular locations. Although current staff and students at a university are located on the university campus, alumni continue to belong to the university community even though they may live and work far afield. The same is true of the bonds between family members. Similarly, while the congregation of a particular church may meet in the church each Sunday, the spiritual body of the Church, made up of all believers, is physically scattered across the globe.[48]

The kingdom of grace, referred to earlier in this chapter, is one such place. All living things act towards final ends or goals, but only some do so in the knowledge that they are acting freely, rationally weighing alternative courses of action before choosing that which they perceive, rightly or wrongly, to be the best. Taken together, these beings comprise what Kant conceived as the 'kingdom of ends', what Leibniz called the 'moral kingdom of grace' or, following Augustine, the 'City of God'. This is a place in which the moral law dictates how these beings ought to regard and act towards each other. In Leibniz's words, they form a kind of society with God at the head. God is 'lord or monarch of minds ... he enters into society with us, as a prince with his subjects; and this consideration is so dear to him that the happy and flourishing state of his empire, which consists in the greatest possible happiness of its inhabitants, becomes the highest of his laws' (*Discourse on Metaphysics* §36: GP IV 462; AG 67).[49]

Members of God's City are expected to promote the greatest possible happiness of all its citizens. For Leibniz, as we've seen, love and justice, the charity or universal benevolence of the wise, provide the means. In loving others, we take pleasure in their happiness. We desire their happiness for their own sakes. Our own pleasure is gained through perceiving the happiness of the other, so that the good of others and our own good come together.[50] The virtuous and wise consider not their own apparent goods, but seek to ascertain the true good and to promote good relations between all rational beings. Acting virtuously, the wise attain a 'serenity of spirit', an 'internal harmony' (*Reflections on the Common Concept of Justice*: L 569–570) that mirrors the external harmony of a society in which the good and perfection of all citizens is paramount.

Taken together, individual places of value, constitute a psychical space that is effectively a spatial version of Leibniz's natural kingdom of final causes, a relational space of value constructed from the perceived and desired goals and ends (the final causes) that govern the conative behaviour of individual living creatures. The kingdom of grace is a subset of this general value space. The City of God forms only part of the natural kingdom of final causes.[51] Beings whose relational qualities – their distinct perceptions and rational volitions – qualify them as members of the kingdom of grace are also members of the wider space that arises from the relational qualities of all living souls and entelechies. Their citizenship of the City of God subjects them to moral scrutiny and appropriate praise and condemnation for acts they self-consciously and freely choose, but it does not follow that such praise and blame should be restricted to actions taken only in relation to other members of the City of God. Adopting the modifications to Leibniz's position proposed earlier, the wise person is expected to consider the perfection of God's creation as a whole, not just the part of it that comprises the kingdom of grace. The wise consider the happiness of fellow human beings, but should also take into account the wellbeing and interests of all living perceiving and conative beings, of all beings making up the space of final causes. In so acting to promote the perfection and goodness and harmonious coexistence of all creatures in this pluralist world, the sage undoubtedly gains immense pleasure from his or her perception of the unfolding beauty of God's creation (*On Wisdom*: GP VII 89; L 428). In practical terms, as a member of the City of God, whenever one finds oneself in any place of value, whether this be the beach, the hospital, the university, the city, the home or the wilderness, it is in our own best interests to cultivate attitudes of love and justice, care and consideration towards and to seek to promote as far as possible the interests and needs of all life forms in that place, not merely the interests of those who will be self-consciously aware that their needs have been met. Leibniz observed that 'the sciences of the just and the useful, that is, of the public good and of their own private good, are mutually tied up in each other, and ... no one can be truly happy in the midst of miserable people' (*Elements of Natural Law*: A VI i 460, L 132). We may amend the latter claim to read: 'no one can be truly happy in the midst of any suffering whatsoever, irrespective of which life forms are affected'.

The relational space and places of value described above reinforce the panpsychist vision of the natural world, not just as a world of physical, moving bodies, but also as a world of living perceiving things – of things with souls whose appetites strive to promote their own wellbeing – only some of whom also belong to the 'smaller' kingdom of grace. The free, self-conscious beings in the latter must assume responsibility both for others in the place of grace, as well as more widely for others in the space of final causes. Space and its component places are formed in part by our perceptions of, and attitudes towards, it. If we regard the natural world of embodied perceiving conative beings with disdain, this perception will be mirrored in the perceptions had by others as they perceive us. Our perceptions and appetitions change our relations to them and them to us and in turn change the places we co-inhabit for better or worse. As humans, we can decide consciously to change our value environments as easily as we change our physical surroundings. The attitudes we adopt towards other living, perceiving, appetitive beings around us alter the tenor, quality and structure of these places and ultimately affect the quality of the value space as a whole. Nor should we forget the nonliving, inorganic aggregate bodies that lack dominant souls or entelechies. They too contribute to the quality of the spaces and places we inhabit, not just as physical quantities but also, either actually or potentially, as objects of beauty from which we take pleasure and as possessors of intrinsic-instrumental relational value towards which we have a responsibility to cultivate respectful and considerate care (chapters 4 and 5). In the following chapter, however, we continue to focus primarily upon relations between living entities as we investigate how Leibniz's doctrines of mirroring and pre-established harmony might be enlisted in the development of a theory of empathic communication between human and nonhuman beings.

Notes

1 See also chapter 5, p. 102.
2 Chapter 1, p. 22.
3 The John described in chapter 2 (p. 42) who perceives the blackbird as part of his world is not the same as the John in some other possible world in which the blackbird plays no role. Similarly, the blackbird whose life experiences include perceptions of John is not the same bird as the blackbird in some other possible world whose complete concept includes no predicates relating it to John. And what holds of John and of the blackbird holds also for all living and nonliving things. All stand in relation to one another and are in part formed by these relations. See also chapter 5, pp. 98–99.
4 See chapter 7, p. 146.
5 See chapter 2, p. 38.
6 A significant proportion of what follows was published in Phemister 2012.
7 Hence, as was also explained there, the perceptions and appetitions had by the soul or entelechy correspond exactly to the motions and resistances of its body, even though there is no actual interaction between souls or entelechies and their bodies. See chapter 2, pp. 40–41.
8 Entelechies that lack self-consciousness also pursue final ends, but being incapable of acknowledging these goals as their own, they are excluded from the moral kingdom of grace.
9 Commentators who read Leibniz as an idealist (e.g. Adams 1994, 225; Rescher 1979) understand spatial relations as founded upon monads' perceptions.
10 Leibniz's view is there sketched only in broad outline. For a detailed exposition and discussion, see Vailati 1997, ch. 4 and Arthur 2014, ch. 7.
11 See chapter 5, pp. 95–96, p. 101 and Leibniz's Fifth Letter to Clarke (GP VII 401; LC 71).
12 See Vailati 1997, 113.
13 On the distinction, see chapter 2, p. 45.
14 *Principles of Nature and of Grace* §2: GP VI 598; AG 207.
15 This is explained in more detail in the section on 'expressive mirroring' in chapter 7.
16 *Discourse on Metaphysics* §9: GP IV 434; AG 42.
17 *Meditations on Knowledge, Truth, and Ideas*: GP IV 422; AG 24. See also chapter 4, pp. 85–86.

18 'Substances have metaphysical matter or primitive passive power insofar as they express something confusedly; active, insofar as they express it distinctly' (*On the Method of Distinguishing Real from Imaginary Phenomena*: GP VII 322; L 365). See also *Monadology* §15 and chapter 2, pp. 37–38.

19 Also see Leibniz's letter to Arnauld, early November 1671: GP I 73; L 150.

20 See also *New Essays*: A VI vi 90, 194; RB 90, 194.

21 *Principles of Nature and of Grace* §18: GP VI 606; AG 213. See also *New Essays*: A VI vi 189; RB 189.

22 Sometimes, however, jealously, fear or shame may prevent us from experiencing fully the pleasure the perception of perfection brings (*On Wisdom*: GP VII 86; L 425). We might add that through, for instance, self-interest or inattention, we may equally fail to perceive another's imperfections and disorder.

23 See chapter 4, pp. 86–87.

24 See also Leibniz's letter to Arnauld, early November 1671: GP I 73; L 150.

25 Recall too Leibniz's description of our reaction to a beautiful painting as an 'image of love' (*Codex Juris Gentium Diplomaticus*, preface: GP III 387; L 422). See chapter 4, p. 85.

26 Leibniz himself admits that some nonhuman animals can reason inductively (*New Essays*: A VI vi 143; RB 143).

27 Leibniz agrees that animals feel affection and anger (*New Essays*: A VI vi 93, 167; RB 93, 167), as well as pleasure and pain, but he thinks their pleasures and pains are less intense than those of rational beings (*Theodicy* §250: GP VI 266; H 281).

28 See also the section on the perfections of living bodies in chapter 7, pp. 141–144.

29 James (2011) defends the view that inanimate things have value in themselves and for this reason should be valued for themselves.

30 *New Essays*: A VI vi 163; RB 163.

31 The qualification of good as 'specific good' denotes a good that is, as Leibniz's contemporary Anne Conway recognized, species-specific (Conway 1996, 32–33; see also Aristotle 2004, VI 7 and X 5). What is good for the sheep is obviously not necessarily also good for the sheepdog. In Leibniz's case, since individuals are themselves *infima species* (chapter 4, p. 75), that is, the lowest species of which they are the sole member, we can understand their 'specific good' as whatever is good for that particular individual in its particular situation.

32 Chapter 2, p. 39.

33 Although the definition of love has been extended so as to allow for love of nonrational creatures, this must not be taken to imply that they are capable of loving us in return. Only those who can recognize the perfection, wellbeing or specific good of others are able to feel love for them.

34 For detailed commentary on Leibniz on justice, see Riley 1996, esp. ch. 4.

35 GP III 386–389; L 421–424.

36 Also see Leibniz's letter to Arnauld, 23 March 1690: GP II 136; L 360.

37 One might discern a hint of paternalism in the notion that the wise know best what is in the interests of others. However, even if this were so, it would not be an authoritarian paternalism. Ultimately the wise desire that all beings that can exercise reason will attain wisdom themselves. Besides, with respect to both rational and nonrational creatures, a degree of humility is required on the part of the wise, given the difficulties of determining what is in their best interests, a process that requires, as we shall see in chapter 7, the exercise not only of reason, but also an imaginative attempt to put oneself in the place of the other.

38 *Codex Juris Gentium Diplomaticus*: GP III 387; L 422; *Elements of Natural Law*: A VI i 461; L 134; *Principles of Nature and of Grace* §§16–17: GP VI 605; AG 212.

39 See also: 'We love God himself above all things because the pleasure which we experience in contemplating the most beautiful being of all is greater than any conceivable joy' (*Elements of Natural Law*: A VI i 461; L 134).

40 '[T]rue piety and even true felicity consist in the love of God ... This kind of love begets that pleasure in good actions which gives relief to virtue, and, relating all to God as to the centre, transports the human to the divine. For in doing one's duty, in obeying reason, one carries out the orders of Supreme Reason. One directs all one's intentions to the common good, which is no other than the glory of God. Thus one finds that there is no greater individual interest than to espouse that of the community, and one gains satisfaction for oneself by taking pleasure in the acquisition of true benefits for men' (*Theodicy*, preface: GP V 27–28; H 51–52).

41 For discussion, see Rateau 2008, 85–91. Sometimes, the good of others may only be attained at some cost to oneself. Recognizing this, Leibniz offers an amended definition of justice as 'the habit of deriving pleasure from an expectation of the good of others, even to the expectation of our own pain', adding nevertheless that 'even though our own pain intervenes, nothing prevents our taking pleasure in an expectation of the good of others' (*Elements of Natural Law*: A VI i 465; L 137).

42 *Monadology* §15: GP VI 609; AG 215.

43 As, for instance, when the cat merely toys with the mouse or bird.

44 When conflicting interests do arise, Leibniz recommends that the needs of the 'better man, that is, the one who loves more generally' are given preference on the ground that he in turn will then be better equipped to help others (to Arnauld, early November 1671: GP I 74; L 150). See chapter 5, p. 107.

45 We may assume a degree of nominalism here: how we choose to carve up space into particular places through language reflects, as Locke recognized, human interests.

46 Leibniz distinguishes societies as equal (for instance, those between friends) or unequal (for instance, between monarch and subjects) and as limited (brought together for particular purposes) or unlimited (being concerned with the 'common good'). See *On Natural Law*: L 429).

47 *Monadology* §21: GP VI 610; AG 216. See also chapter 4, note 11 (p. 89).

48 Leibniz himself identifies the 'church of God' as the sixth and final 'natural society', the others being constituted by relations of friendship, cooperation, education and obligation between husband and wife, parents and children, master and servant, the whole household and civil society (*On Natural Law*: L 428–429).

49 Here too, Leibniz insisted, the laws of justice, divinely administered, ensure that all good acts are rewarded and evil punished (*Principles of Nature and of Grace* §15: GP VI 605; AG 212).

50 See, for instance, *Remarks on the Three Volumes Entitled Characteristics of Men, Manners, Opinions, Times ...* (GP III 425; L 630) where Leibniz asserts that reason commands us to promote our own happiness; that our own happiness is best promoted in turn by acting virtuously (here understood as 'reasonably'); and that to act thus involves disinterestedly seeking the good of all.

51 See above, p. 113.

7   Expressive communication and empathy

Can I see another's woe,
And not be in sorrow too?
Can I see another's grief,
And not seek for kind relief?
                    ('On Another's Sorrow', from William Blake, *Songs of Innocence*)

In chapter 5, we explored the relational identities of the individuals that comprise the Leibnizian world and developed a theory of relational value under which the internal worth of individual, living and nonliving, things is conceived in terms of their external contribution to others' wellbeing and the perfection of the whole. In chapter 6, we expanded upon these ideas to propose an extension of Leibniz's account of relational physical space to embrace a conception of relational psychical space arising from the qualitative value relations between individual entities, highlighting the contribution of compassionate and loving feelings towards others to the life-enhancing features of space and places. In this chapter, we explore relations between living things from the perspective of Leibniz's doctrine of mirroring or mutual expression, using this to sketch in broad outline a theory of empathy as an underpinning mechanism of just, harmonious loving relations with all living beings. We shall continue to regard relations between individuals primarily, though not exclusively, as relations of presently coexisting or contemporaneous living beings. In chapter 8, however, we shall turn to the issue of temporal value relations between living things and to consideration of the influence of the past and the present on the future.

Communication

The lover takes pleasure from seeing the perfection and wellbeing of the beloved and behaves in ways that help the beloved to live well and to flourish. But to do this, the lover needs to know what the beloved needs in order to flourish. The beloved must be able to communicate their needs to those who love them. Philosophers such as John Locke considered communication primarily in terms of the communication of ideas by humans through the spoken or written word. Thus, Locke opens Book 3 of his *Essay concerning Human*

Understanding ('On Words') with the suggestion that the purpose of the peculiarly human ability to make articulate sounds signs of ideas in our minds is solely in order that our ideas 'might be made known to others, and the Thoughts of Men's Minds be conveyed from one to another' (*Essay* 3.1.2). Locke does admit that some nonhuman animals have ideas and memories. For instance, songbirds, he contends, must have ideas of notes and tunes in order to be able to reproduce these as they do (*Essay* 2.10.10). However, even the songbird, he believes, cannot turn these sounds into representative signs of its ideas in any meaningful way. The bird copies the sounds but does not translate its remembered ideas into signs that have meaning or signification. Animals can '*frame articulate Sounds*' (*Essay* 3.1.1) but they cannot use these sounds 'as *Signs of internal Conceptions*' (*Essay* 3.1.2). The implication we may draw from Locke's remarks is that birds and other nonhuman living beings do not communicate ideas to others through the use of signifying sounds. This echoes similar thoughts found in Descartes's writings. For Descartes,

> there are no men so dull-witted or stupid – and this includes even madmen – that they are incapable of arranging various words together and forming an utterance from them in order to make their thoughts understood; whereas there is no other animal, however perfect and well-endowed it may be, that can do the like. This does not happen because they lack the necessary organs, for we see that magpies and parrots can utter words as we do, and yet they cannot speak as we do; that is, they cannot show that they are thinking what they are saying.
>
> (*Discourse on the Method*: AT VI 57; CSM I 140)

Renaissance thinkers had not been so anxious to rule out the linguistic abilities of nonhuman animals. In *An Apology for Raymond Sebond,* the sixteenth-century humanist and essayist, Michel de Montaigne attributes the power of speech to dogs, horses, partridges and other nonhuman creatures.[1] Nowadays, it is generally accepted that many living beings, such as mammals, birds and insects, communicate with each other using meaningful sounds, but whether these sounds constitute languages, understood as complex syntactical and semantic rule-governed systems, is far from clear. Nevertheless, even if nonhuman creatures lack linguistic abilities, they do appear capable of communicating with each other by other means. Their needs and interests can be communicated in ways other than by ideas signified by words. Whales sing, bees dance, ants emit pheromones.[2] Scientific research into the details and extent of animal communication is in its infancy, but it is already clear many nonhuman creatures use sounds, smells, gestures and other strategies to convey information such as location and group affiliations, sources of food, potential danger and sexual attraction. There are also channels of communication across species boundaries, including between animals and humans. Those in close relationships with companion animals are well aware of this. When the dog drops the lead into his owner's lap she knows perfectly well that this is the dog's way of saying he wants her to take him for a walk or that when he places his paw on her lap, he wants her attention, which once secured, he'll let her know why. With growing familiarity, she finds herself tuning into the nuances of the dog's barks as they signal excitement, pain, fear, shock, anger, belligerence, assertiveness and so forth. The dog in turn is attuned to her as she communicates with him by sound and gesture. Even the mere raising of an eyebrow can be sufficient to signal an intention to go for a walk or express disapproval over something the dog has done or is clearly poised to do. Human-to-human communication also makes extensive use of gestures and other nonlinguistic communication tools. Emotions are powerfully expressed through music and other art forms. Facial expressions and physical postures communicate joy, love, pity, sympathy, trust, honesty or deceit, disdain, dislike, and many other affects. Both anger and tenderness are communicable by touch; both fear and attraction by smell.

Leibniz uses the term 'communication' in two different and opposed senses. The first is common throughout his corpus and bears the closer resemblance to our ordinary understanding of the term. In this sense, communication is conceived as the transference of something from one thing to another, as, for instance, the transmission of physical matter or force or the transmission of psychical data or information from one substance to another. We may call this, 'communication-by-transference'. Leibniz consistently denies that this type of communication, in any of its forms, occurs in the universe. We find him rejecting communication-by-transference in the *New System of the Nature and of the Communication of Substances* that he published anonymously in two consecutive parts in 1695 in the 27 June and 4 July issues of the *Journal des Savants* (WF 7). In the first part, Leibniz outlines the nature of substances as indivisible, self-contained, active units of force. In the second part, he addresses the question of how such indivisible substances might communicate with each other even though 'it is impossible that the soul or any other true substance should receive anything from outside' (GP IV 484; WF 17). The 'windowlessness' of substances is incompatible with the notion of communication as involving the transference of something between the communicants: as indivisible, monads have no parts that would allow them to be

> altered or changed internally by some other creature, since one cannot transpose anything in it, nor can one conceive of any internal motion that can be excited, directed, augmented, or diminished within it, as can be done in composites, where there can be change among the parts.
>
> (*Monadology* §7: GP VI 607; AG 213)

However, although Leibniz rejects 'communication-by-transference', another sense of 'communication', which we may call 'communication-by-agreement' can be found in his works. This use of the term 'communication' occurs in clarifications of the *New System* penned by Leibniz in the years immediately following the 1695 articles. The *First Explanation*, [3] published in the 2 and 9 April 1696 issues of the *Journal des Savants*, was written as a reply to Simon

Foucher who had published critical remarks on Leibniz's *New System* article in the 12 September 1695 issue of the same journal.[4] In his reply, Leibniz explains that his 'theory of harmony or concomitance' that proposes an 'exact correspondence between substances' was deliberately intended to 'explain the communication between substances' without resorting to an unintelligible '*transmission* of species'[5] from one substance to the other (GP IV 494–495; WF 48–49). Rather than denying any kind of communication between substances, Leibniz here offers the harmony or agreement between substances as an alternative explanation of the phenomenon.[6] A decade earlier, in a remark that anticipates his claim in the *New System* that the agreement between substances stems from their having a 'common cause',[7] Leibniz had already observed that 'God alone brings about the connection and *communication* among substances, and it is through him that the phenomena of any substance meet and agree with those of others and consequently, that there is reality in our perceptions' (*Discourse on Metaphysics* §32; GP IV 458; AG 64, my emphasis). Here, as in the *First Explanation*, he confirms that the agreement between substances counts as communication, even though it is executed in a manner that does not require the transmission of anything from one substance to the others.

In the *New System* Leibniz had also raised the question of how the soul and the body communicate with each other. Here, too, there is no physical transmission that would constitute the ordinary understanding of 'communication'. The 'mutual relationship' of substances with each other, he claims, 'alone constitutes *the union of mind and body*', ensuring that the animal spirits and the blood in the body assume 'exactly at the right moment the motions which correspond to the passions and perceptions of the soul' (GP IV 484–485; WF 18). Nothing passes from the soul to its body or from the body to the soul. Each acts only in accordance with its prearranged plan. In the absence of any transfer of 'species or qualities' between substances, the union of the soul with its body consists only in the agreement of the soul and the particular substances that comprise its body (GP IV 484; WF 18). In the *New System*, Leibniz resisted the inclination to describe the agreement or union of soul and body as an instance of 'communication' between the soul and body. However, shortly after publication, he composed a summary of the journal article and sent it to Henri Basnage de Beauval, editor of the *Histoire des Ouvrages des Savants*. Basnage must have responded, for Leibniz followed up with a response and it is in this follow-up note that he writes of the 'communication or harmony' between the soul and the body (to Basnage, 3 January 1696: GP IV 498–500; AG 147, WF 62),[8] clearly happy to describe the harmonious relation between the soul and the body as a communicative relation, that is, 'communication-by-agreement'. Fourteen years later, in Part 1 of the *Theodicy*, Leibniz describes the pre-established harmony between the soul and the body as a kind of '*metaphysical*' (as opposed to a '*physical*') communication by which the soul and the body together 'compose one and the same *suppositum*, or what is called a person' (§59: GP VI 135; H 155). We may understand this in line with the Spinozist account of the union offered in chapter 2 whereby both the soul's perceptions and its body's motion and resistance are two different kinds of modifications that result from the same primitive forces that constitute the essence of the individual substance.[9] What is interesting for our present purposes is that the union of soul and body is a communication based on harmony and agreement between perceptions in the soul and motions in the body that does not involve any interaction between soul and body or the transference of anything from the one to the other.

Etymological roots of this Leibnizian use of the term 'communication' as 'communication-by-agreement' or, what is effectively the same thing, 'communication-as-harmony' can be traced back to late thirteenth- and early fourteenth-century Old French. According to the *Oxford English Dictionary*, the term served then to indicate only the 'fact of having something in common with another person or thing' (*OED*, 'communication, n.' etymology). The same root underpins the closely affiliated notions of 'communion' and 'community'.[10] This older sense of 'communication' allows for the possibility that things can 'have something in common' without there being any transmission or transference of anything from one to the other. And such is the case with communication between Leibniz's individual substances. Being 'windowless', nothing is actually transferred from one substance to another, but they nevertheless share a great deal in common with each other. Indeed, the mutual expression or mirroring of all by each ensures that they have the whole world in common as each individual substance perceives the world in its entirety, differentiated from the others only insofar as this world is perceived from its own perspective or point of view. There is comm*union* or sharing between them insofar as each contains the whole; together they form a comm*unity* of interconnected beings, a world of harmoniously coexisting beings.

Expressive mirroring

I have been suggesting that all Leibnizian creatures communicate with each other and that they do so through their constant expressive mirroring or perception of each other's states, constituting their 'communication-by-agreement'. Leibniz uses 'expression' as a general term covering multiple examples of representation, from maps that represent the contours of the land to the representation of thoughts by ideas or the expression or representation of feelings and emotions by works of art.[11] Substances' perceptions or perceptual states[12] also count as 'expressions': 'Expression … is a genus of which natural perception, animal sensation and intellectual knowledge are species' (to Arnauld, 9 October 1687: GP II 112; LA 144).

Monads' perceptions are complete representations (from each monad's particular perspective) of the current state of the world from moment to moment. Each perception or perceptual state is a 'multiplicity in unity' comprising many simultaneous perceptions, each of which is itself a perceptual state composed of simultaneous perceptions, ad infinitum. In an early text, Leibniz described the confused perception of a sensible quality (such as the perception of an object's colour) as 'not one perception, but an aggregate of infinitely many perceptions'

(*On Forms, or, the Attributes of God*, Paris Notes, April 1676: DSR 70–71). In the 1684 *Meditations on Knowledge, Truth, and Ideas*, whose thesis Leibniz continued to endorse well into his later philosophy, he explains how, when we perceive a colour, such as green, we have only,

> a perception of figures and motions, but of figures and motions so complex and minute that our mind in its present state is incapable of observing each distinctly and therefore fails to notice that its perception is compounded of single perceptions of exceedingly small figures and motions. So when we mix yellow and blue powders and perceive a green colour, we are in fact sensing nothing but yellow and blue thoroughly mixed; but we do not notice this and so assume some new nature instead.
>
> (*Meditations on Knowledge, Truth, and Ideas*: GP IV 426; L 294)[13]

Indeed, *all* perceptions, irrespective of their various degrees of obscurity, confusion, clarity or distinctness, involve the infinite.[14] All are composed of infinitely many smaller perceptions. As Leibniz explained to Bayle, both imagination as well as intellect are required in order for us to think and to reason abstractly and the 'confused thoughts (which invariably accompany the most distinct that we can have) ... always involve the infinite, and not only what happens in our body but also, by means of it, what happens elsewhere' (*Reply to the Comments on the 2nd Edition of M. Bayle's 'Critical Dictionary'*: GP IV 563–564; WF 117).

As this last statement makes clear, Leibniz recognized not only that every perceptual state in a living thing is an exact expression of the state of the entire universe at that moment, but also that it is perceived through the prism of the soul's own body. The soul's perceptions,

> correspond by themselves to what happens in the whole universe. But they correspond more particularly and more perfectly to what happens in the body assigned to it, because the soul expresses the state of the universe in some way and for some time, according to the relation other bodies have to its own body.
>
> (*Discourse on Metaphysics* §33: GP IV 458; AG 64–65)[15]

Perhaps the fullest account is that which Leibniz gave in his 9 October 1687 letter to Arnauld in which he describes how the multiplicity that constitutes the soul's infinitely divided body is represented in the unity that is the soul's current perceptual state and explains how through its expression of the current state of its own body, the soul actually perceives the whole universe. The passage is worth quoting at length:

> In natural perception and in sensation, it is enough for what is divisible and material and dispersed into many entities to be expressed or represented in a single indivisible entity or in substance which is endowed with genuine unity. One cannot doubt the possibility of a noble representation of many

things in a single one, since our soul provides us with an example of it. But this representation is accompanied by consciousness in the rational soul, and then it is called thought. Now, this expression occurs everywhere, because every substance is in harmony with every other and undergoes some proportionate change which corresponds to the smallest change occurring in the whole universe, although this change is more or less noticeable to the extent that other bodies or their actions have more or less connexion with ours. I believe that M. Descartes himself would have agreed with this, for he would undoubtedly grant that because of the continuity and divisibility of all matter the smallest movement extends its effect over neighbouring bodies and consequently from neighbour to neighbour *ad infinitum*, but proportionately decreased; thus our body must be affected in a way by the changes of all the others. Now to all the movements of our body there correspond certain more or less confused perceptions or thoughts of our soul, therefore the soul too will have some thought about all the movements of the universe; and in my view every other soul or substance will have some perception or expression of them.

(GP II 112–113; LA 144)

On this model, monads' perceptions fulfil the criterion specified in Leibniz's technical definition of 'expression', where the 'particulars' are the component perceptions in the soul's perceptual state and the figures and motions of the body's parts, respectively:

it is sufficient for the expression of one thing in another that there should be a certain constant relational law, by which particulars in the one can be referred to corresponding particulars in the other.

(*Metaphysical Consequences of the Principle of Reason*:
C 15; PW 176–177)[16]

Each tiny perception that goes to make up the soul's complete perceptual state at any one moment maps isomorphically[17] to each and every physical point of the soul's organic body at that moment, registering, albeit for the most part unconsciously,[18] each and every change that is wrought on the body as it interacts with those around it, and through the interconnectedness of bodies in the material continuum, indirectly registering every change in the whole universe.

Moreover, because each embodied soul expresses its own body in this way, it follows that there is a harmonious expressive ordering of all souls' or entelechies' perceptions as well. The changes we perceive in our own bodies have been effected through interactions with bodies that others perceive as their own. From their side too, our interactions with external bodies bring about changes that others' perceive in their own bodies. We mirror the changes that others perceive in their bodies through the changes that their bodies make on our own, and vice versa. Thus, each soul or entelechy not only expresses its own and others' bodies; it also expresses the other perceiving souls and entelechies that in turn express their own and others' bodies. In this way, the 'communication or harmony' of the soul with its body, which Leibniz described to Basnage, is also a 'communication or harmony' of one soul or entelechy with other souls and entelechies, such that it is in theory possible to map the perceptual experiences of one substance to the experiences had by all the others. Even without the transfer of anything from one substance to another, the pre-established order between souls and bodies ensures that there is also communicative harmony or 'communication-by-agreement' between all living perceiving beings. The representative natures of the Leibnizian souls are communicative natures.

How might communication-by-agreement between the mind and the body assist us in determining the needs and interests of others? Can it help us ascertain the most efficient means of improving the lives and experiences of both humans and nonhuman creatures alike? From the empirical evidence of the states of their bodies, we assess whether plants, fish, animals and other living things are sick or healthy. By paying attention to their physical appearance and behaviour, we can determine what constitutes their physical health and wellbeing and can determine what actions on our part will best promote what Leibniz has called the 'perfection of their bodies'. We have no direct access, from the 'inside' so to speak, to the minds and souls of other living beings, but we commonly take the external states of their bodies to be reliable indicators of the internal states of their souls or entelechies. For instance, from a person's demeanour and gestures, we discern whether they are angry, happy or sad; from a pet's behaviour, whether it is at ease or in pain, content or fearful.[19] Anecdotal evidence suggests that many pet owners believe their pets are in turn aware of their owner's emotions and feelings.

The harmonious communication-by-agreement between substances and between souls and their bodies explained above gives metaphysical underpinning to these natural inclinations to infer psychical states from physical states. We know that Leibniz promoted the natural embodiment of all souls and entelechies and the corresponding natural ensoulment of all organic (or organized) living bodies. And we have seen how each state of the soul is correlated exactly with the corresponding state of its body so that the soul is, in effect, an exact mental image or representation of the physical state of the body and the body in turn is an exact physical expression of the perceiving, appetitive soul.[20] The features of souls are not the features of bodies, but because souls and bodies are exact expressions of each other, inferences from the health of the one to the health of the other are justified. Harmonious communication allows us to read the state of the mind or soul from the appearance of the body. Given the union of the soul with its body, we can be assured that if the body is healthy, so too the perceptions in the creature's soul or entelechy will possess positive characteristics akin to those that we know in ourselves as pleasure and that when it is sick, the corresponding perceptions will veer towards pain. The closer the biological relationship to ourselves, the more easily we can judge the affective states of others. We may be relatively confident in our assessment of the psychical states of apes, horses, dogs, cats, foxes and other mammals. However, getting a sense of the psychical health of fish, insects, trees, plants, bacteria and other life forms is far more challenging. To understand Leibniz's views on these matters more thoroughly, we examine in the next couple of sections the ways in which the 'perfections of the body' might serve as useful indicators of the 'perfections of the soul'.[21] We begin with the perfections of the soul.

The perfections of the entelechy or soul

In the opening paragraphs of the preface to his *Theodicy*, Leibniz says of human souls that their perfections are the same as the perfections of God, which they possess to a lesser degree:

> The perfections of God are those of our souls, but he possesses them in boundless measure; he is an Ocean, whereof to us only drops have been granted; there is in us some power, some knowledge, some goodness, but in God they are all in their entirety. Order, proportions, harmony delight us; painting and music are samples of these: God is all order; he always keeps truth of proportions, he makes universal harmony; all beauty is an effusion of his rays.
>
> (*Theodicy*, preface: GP VI 27; H 51)

Lesser versions of God's omnipotence, omniscience and wise benevolence are present in human souls. A semblance of God's omnipotence is evident in the soul's primitive active power and its (sometimes free) spontaneous activity manifested by the internal movement or appetition from one perception to the next. God's omniscience is echoed in the human soul as its distinct perceptions that, depending on the degree of distinctness attained (itself dependent on the strength of appetitive force that brings the perception to light), constitute knowledge of the true good over the merely apparently good. And a likeness of God's benevolence is found in the human soul as its virtue or moral goodness, grounded in the volition or rational appetite to bring about what the soul has distinctly perceived to be truly good.

For Leibniz, the wellbeing or flourishing of the self-conscious, rational soul resides in the perfecting of moral character, piety and virtue: 'the most perfect of all [created] beings ... are minds, whose perfections consist in their virtues' (*Discourse on Metaphysics* §5: GP IV 430; AG 38). As we found in chapter 6, true wisdom combines knowledge of goodness, perfection, order and beauty with the rational will to act benevolently and with justice and charity. The wise attain the freedom, virtue and happiness that arise naturally from the appreciation of God's perfections and the beauty and harmony of God's creation. God possesses all perfections to the highest degree. Minds possess these perfections in lesser degrees, but nonetheless, their rational appetitions are strong enough to move them towards clear and distinct perceptions that allow them to differentiate, to some extent, the true goods from the merely apparent goods. Moreover the perceptions themselves, when sufficiently distinct, can distinguish within themselves the content of the perception and perceiver's own act of perceiving.[22] In this way, the rational, distinctly perceiving soul becomes conscious of itself. It becomes aware of itself as an 'I' that is distinguished from other perceived 'I's. It perceives itself as an active being, the free initiator of its actions and thereby also responsible for the consequences of its actions. It realizes, in effect, that it is a moral agent. Meanwhile, Leibniz holds that every being naturally desires that which it perceives as being good. In God, this is manifested as the will to create whatever is distinctly perceived as truly good: the best possible and most harmoniously ordered world. God's omniscience ensures God perceives the true good; his benevolence ensures that he wills to create it; and his omnipotence ensures that he has the power to do so. Hence, God's volitions are always directed towards what is truly the best and are always efficacious. Insofar as the rational appetites or volitions of finite creatures are founded upon distinct perceptions of the true good, they accord with divine teleology. The distinctly perceiving wise person wills what God wills.

God's omnipotence, omniscience and wise benevolence operate in unison, ensuring that he knows not only which world is the best, but also that he has the will and the power to create it. In so doing, God creates that possible world that is the most ordered, harmonious and varied. But God not only creates harmony, variety and order; he himself possesses the perfections of the world he creates.[23] God himself, Leibniz had stated in the passage quoted above, 'is all order; he always keeps truth of proportions, he makes universal harmony; all beauty is an effusion of his rays' (*Theodicy*, preface: GP VI 27; H 51). Leibniz believes that when any perfection is perceived,

> such as understanding, courage, and especially beauty in another human being, or in an animal or even in a lifeless creation, a painting or a work of craftsmanship, as well. For the image of such perfection in others, impressed upon us, causes some of this perfection to be implanted and aroused within ourselves.
>
> (*On Wisdom*: GP VII 86; L 425)[24]

The same presumably holds for the perceptions of the perfection and beauty of the world as a whole and of the perfection, order and beauty of the God who created it. By reflecting the beautiful, ordered and varied harmony of the world and the perfect order and beauty of God, the rational soul itself is a thing of beauty and harmonious order. When the soul perceives things of beauty and harmony, the representational content of its perceptions is beautiful and harmonious. Since its perceptions are the soul's internal qualities, the perceiving soul too acquires the same qualities. That is to say, when the soul's internal qualities are ordered and beautiful, so too is the soul itself. The more distinctly a mind perceives this beauty and order, the greater its own perfection:

> since each distinct perception of the soul includes an infinity of confused perceptions which embrace the whole universe, the soul itself knows the

things it perceives only so far as it has distinct and heightened [*revelées*] perceptions; and it has perfection to the extent that it has distinct perceptions.

(*Principles of Nature and of Grace* §13: GP VI 604; AG 211)

God's perfections are not replicated only in human or rational souls. All the living creatures in Leibniz's world, from very large to the infinitely small, have power, perception and appetite, corresponding to the divine perfections of omnipotence, omniscience and benevolent will. Just as do our distinct perceptions, others' perceptions of God's creation also replicate the perfection of divine order. All souls and entelechies perceive, however obscurely or confusedly, the perfectly harmonious created world. This at least would seem to be the implication of Leibniz's remark in the *Principles of Nature and of Grace* that '[o]ne could know the beauty of the universe in each soul, if one could unfold all its folds' (§13: GP VI 604; AG 211). We have already noted[25] Leibniz's admission that 'every substance bears in some way the character of God's infinite wisdom and omnipotence and imitates him as much as it is capable' (*Discourse on Metaphysics* §9: GP IV 434; AG 42). All created substances – all 'derivative substances' – contain the perfections that God possesses 'eminently' (*Principles of Nature and of Grace* §9: GP VI 602; AG 210).[26] Each soul or entelechy has a spark of God's wisdom insofar as each 'expresses, however confusedly, everything that happens in the universe, whether past, present, or future'. So too, each substance has a power akin to God's omnipotence, insofar as they 'accommodate themselves' to each other (*Discourse on Metaphysics* §9: GP IV 434; AG 42). The tree outside my window has power over me insofar as I cannot help but perceive it when I look in its direction. Others dictate what I perceive as much as I dictate what they perceive. In addition, all created substances share in God's omnipotence in being active forces. All act from their own internal spontaneity, free from any external force. All too have some semblance of the divine perfect benevolent will towards what is truly good, for the appetitions of finite beings unfold the series of perceptions or experiences (sensations, feelings, emotions) that constitute their lives as lived and take each in the direction of what it perceives as being good. Lacking God's distinct perceptions of the true goods, they may often mistakenly strive for false goods, as for instance, when an insect is attracted to the scent emitted from the Venus flytrap (*Dionaea muscipula*), but their wellbeing is furthered when what is sought is that which is truly in their best interests (even if they do not distinctly perceive it as such) and when these good desires are satisfied.

To varying degrees, therefore, souls and entelechies share the perfections of God.[27] Primarily power or force, perception or knowledge, appetition or will, to these may also be added the order and beauty they share with God through their representations of the ordered harmony of the best possible world.[28] We turn now to the question of the perfections of bodies and to consideration of how these might serve as physical and empirically accessible indicators of the perfections of souls.

The perfections of the body

In his *Essay concerning Human Understanding*, John Locke had raised the possibility that matter might be endowed with the capacity to think (*Essay* 4.3.6). In a letter to Locke's acquaintance, Thomas Burnett of Kemnay, Leibniz notes his disapproval of the very notion of 'thinking matter'. This is just as we would expect. For Leibniz, thought or distinct perception is a perfection of the rational soul. Bodies have motion and resistance, but they do not perceive or feel. Nevertheless, in the same letter to Burnett, Leibniz reports approvingly of other perfections that Locke had attributed to bodies. Taking his lead from Locke's argument in his Third Letter to his critic, the Bishop of Worcester, Edward Stillingfleet (Locke 2008, 491–496), Leibniz reads Locke as claiming that 'matter ... can receive from God different degrees of perfection, like force and motion; life and vegetation; sensation and spontaneous motion; and perhaps even reason and thought' (to Thomas Burnett, end (?) 1699–early 1700: A VI vi 31–32).[29] Apart from reason, thought and, presumably, also 'sensation', since this too is a quality peculiar to souls,[30] Leibniz declares himself in 'complete agreement' with the English philosopher.

All of the remaining Lockean bodily perfections – force, spontaneous motion, life and vegetation (understanding this as growth or development) – are to be found in organic bodies, conceived by Leibniz as living organized machines endowed with souls or entelechies. Always ensouled, always dominated by a unifying soul or entelechy, organic bodies are aggregates of monad–body unities or 'corporeal substances'.[31] As such, they are preformed living machines with infinitely nested complex organizational structures that spontaneously, through the mechanical operation of their derivative active and passive forces, generate the motions and resistances required in their parts to enable them to grow and to maintain their existence, repairing themselves when necessary and in due course reproducing more of their kind.[32]

Divine omnipotence is reflected in souls and entelechies as their primitive active force. Corresponding to souls' active forces are their bodies' derivative active forces. Thus, the semblance of divine omnipotence that is found in souls and entelechies is also found in bodies as their derivative active force. In similar fashion, divine benevolence or will is reflected in souls and entelechies as their appetitions. These are the fleeting modifications of souls' active forces that take them from one perception to the next. Corresponding to these changes to souls' perceptual states are the motions of their bodies – the manifestation of bodies' derivative active forces – that take each body from one position in space to the next.[33] Thus, just as primitive active force in the soul results in the sequence of appetitions that move the soul from one perception to the next, so too the body's derivative active force gives rise to the series of motions that take the body from one place to the next. In this way, the motion of its organic body is the means by which the dominant soul brings to fruition that for which its appetitions strive[34] and the divine benevolence or will that is mirrored in souls' appetitions is mirrored also in bodies through their motion.

However, what of divine omniscience? This perfection is shared by minds and souls through their knowledge and perception. But what corresponds to it in the body? If the body's derivative active force corresponds to the soul's primitive active force and the body's motions correspond to the soul's appetitions, we would expect that there would be something in bodies that corresponds to the soul's perceptions. Of the Lockean bodily perfections, the remaining candidate is the body's 'life or vegetation'. Now the concepts of life or vegetation (growth or development) might at first appear to be too closely associated with the notions of force and motion for them to be correlated with the souls' perceptions rather than with the soul's appetitions. However, a key feature of Leibniz's living bodies is their organized structure and, as we noted earlier, organization or order is also a key feature of souls' perceptions. Living bodies are organic bodies, *organized* machines. Moreover, the integrated organization of the parts of living bodies is essential to their biological functioning. Just as in a wound-up mechanical clock the cogs and wheels of its mechanism must be in good working order if the force present in the springs is to be able to move the hands around the clock face, so too the mechanism of the living body must be in good working order if the body's derivative force is to move the parts of the body in the manner required for the body to perform the physiological functions that successfully maintain its life and that allow it to develop through growth.

In a postscript to one of his letters to Sophie, Electress of Hanover, Leibniz clarifies how in his view 'all the impressions [in a soul] can be traced, so to speak, in the infinite varieties of shapes and motions that there are in the surrounding matter, which preserve something of all preceding effects' (to Sophie, 6 February 1706: GP VII 570; LS 351). Leibniz had made essentially the same claim in his 9 October 1687 letter to Arnauld, quoted at length earlier in this chapter.[35] On that occasion, the situation was reversed: in the letter to Arnauld, it is the soul that is said to trace or mirror (i.e. to perceive or express) all the movements of its body. In so doing, the soul expresses the infinitely many effects made upon its own body as it interacts with other bodies. In the body itself these effects are distributed between its infinitely many parts, but in perceiving them, the soul holds them all in one single unifying and indivisible perceptual state: 'what is divisible and material and dispersed into many entities [the organic body] is ... expressed or represented in a single indivisible entity or in substance which is endowed with genuine unity' (GP II 112; LA 144). The present state of a living body is the combined effect of all the preceding causes that have led to each of its parts being in the precise state it is. And because of the uninterrupted continuity and interconnection of bodies, these causes encompass everything that has happened in the universe from its beginning to the present moment (GP II 112–113; LA 144). Had anything in the past been different, the present would not be as it is now. Correlatively, everything that has actually happened in the past has played a role in shaping the present.[36] Consequently, if one had a complete understanding of the present state of any individual organic body, it would be possible to trace the entire history of the universe that has led up to the present moment. Not only does every effect depend upon 'an infinite number of causes', but also 'every cause has an infinite number of effects' (to Arnauld, 30 April 1687: GP II 98; LA 123). Any body endowed with the unity conferred upon it by its dominant entelechy, i.e. any organic body, is such a cause.[37] Each gives rise to an infinite number of effects. Given its role as both the effect of past causes and the cause of future effects, we can understand how it comes to be that 'the lineaments [*les traits*] of the future are formed in advance and ... the indications [*les traces*] of the past are preserved for ever in each thing' (to Arnauld, 30 April 1687: GP II 98; LA 123). Each organic body contains traces of the past (its causes), its present state, and indications of its own and others' future states (its effects). In the present context, the important point to note is that this means that there is a sense in which, just like its dominant soul, each organic body possesses the divine perfection of order, for each contains the order of the universe in itself. Each organized living body at every moment replicates the order of the universe and mirrors the perfect orderliness of God's omniscience as much as does each mind, soul or entelechy in its perceptions.

Given these correspondences between the perfections of the body (derivative active force, motion, organizational structure) and the perfections of the soul (primitive active force, appetitions, perception), what do the perfections – or lack thereof – of particular organic bodies tell us about the states of their dominant souls or entelechies? Clearly, low levels of derivative active force (or correspondingly high levels of primitive passive force) in the body limit its ability to move. High levels of energy are indicated by motion and activity, but we also know that, generally speaking, active bodies are healthy bodies. We know in our own cases that we tend to have high levels of energy when we are in good health. Physically, our bodies exude vitality; our skin glows; and our bodily organs operate efficiently. Psychically, we find ourselves filled with energy to plan and execute projects with success. There is excitement, joy and pleasure, enthusiasm and creativity. Our appetitive force moves us from one perception to the next with ease and the perceptions attained are clear and distinct, making us attentive to our surroundings, aware of what's going on. On the other hand, when our bodies are sick or diseased, when their mechanical organization is disrupted, their energies are depleted. Their movement becomes slow and sluggish. We may find we cannot even raise our bodies to an upright position. There may be problems breathing, digestive complications, inability to regulate body temperature, and so forth. All these are indications that the body's internal mechanisms are in need of repair and that the body is finding it difficult to self-repair. On the psychical side, disease or sickness in the body is matched by sluggish and confused thoughts, a slowness in moving from one idea to the next, lack of concentration, sleepiness, inability to complete projects, lack of creative energy, depression, lack of enthusiasm and joy, feelings of pain in the body and mind, and so forth.[38] The links between the perfections of the body and those of the soul are obvious to us as humans. From our own experiences, we know both sides of the equation, inside and out. At one and the same time, we know both the current states of our own souls as well as the current states of our bodies. We feel our energy simultaneously in our minds and in our bodies. When we walk down a country lane, we experience the correlation between our appetitive transition from one perception to the next and the movement of our bodies, appreciating the changes in our perceptions as they monitor the changes in our relations to our environment, one moment smelling the newly cut hay, the next moment enjoying the scent of the wild garlic by the side of the path. We feel the elevation and invigoration of our minds as much as that of our bodies as we climb the hill or mountain.

With other human and nonhuman beings, we can only conjecture their psychical states from their physical appearances. However, the principles are the same. In each living thing, the health of the organic body can be taken as indicative of the overall wellbeing of the creature. Given the correlations between the perfections of the body and those of the soul, the perfections we are able to discern in the bodies of human and nonhuman beings may be regarded as guides to the correspondingly pleasant nature of the experiences enjoyed by the body's dominant soul or entelechy. Without fully adequate knowledge of bodies, these signs are always susceptible to doubt.[39] Nevertheless, the excitedly buzzing pollen-collecting bee, the salmon leaping upriver to the spawning sites of their birth, the vigorously growing bramble bush, the profusion of flowers on the rhododendron, the lichen spreading over the stones and fallen wood are well-organized bodies whose activity suggests the presence of strong derivative active forces that have given rise to such energetic motion. Accordingly, we presume too that they experience good psychical as well as physical health. From the perspective of the Leibnizian panpsychist, it is reasonable to believe that the experiences had by salmon tight-packed into pens in fish farms are far less pleasant than they would be if the salmon were free to swim in the open seas and to return to freshwater when the time comes to spawn; that the perceptions had by the gull covered by oil from a spill on the beach are more painful than those had by its cousin still free to fly in open air; and that those of the fruit fly in the laboratory are more stressful and less pleasant than they would be were it free to roam wherever it pleased. It is reasonable to presume too that the experiences of the bee that attains the clear perceptions of pollen for which its appetitions strive, of the salmon whose soul gains sight of the spawning ground it seeks, and of the plant whose appetitions strive towards experiences of its body growing and reproducing are each better and more pleasant than the experiences had by another bee, salmon or plant whose appetitions have either gone in less healthy directions or have not been backed by sufficient primitive force to reach their target.

Empathy and communication

Reading the needs of other living beings from their external appearances does not necessarily also prompt us to want to promote their interests. This is especially so if the others' perceived needs appear to conflict with our own and particularly with respect to nonhuman beings, there is often a temptation to downplay their interests and to present their needs as somehow less significant than our own. After all, it might be said, if salmon and gulls are not rational thinkers, their perceptions will never reach the degrees of distinctness that support self-conscious thought. Their metaphysically primitive and physically derivative forces are sufficient to move them through the air and sea and the degree of distinctness in their perceptions clear enough to locate their food, but their perceptions do not need to be so distinct as to allow them to recognize themselves as perceivers distinct from other perceivers, to see themselves as moral agents. Presumably, too, with lower degrees of force, their perceptions will never attain the higher levels of pleasure that come with the contemplation of beauty and of God. But nor will they feel the despair and anguish that comes from knowledge that they have not acted as well as they might. Their physical pain too may, if we are so inclined, be dismissed as relatively insignificant – *only* a semi-pain, a semi-perception, *only* a bare feeling. Leibniz himself was prone to such reasoning. While reluctant to deny animal feeling, declaring that 'one cannot reasonably doubt the existence of pain among animals', he nevertheless appears immediately to diminish its relevance: 'but it seems as if their pleasures and their pains are not so keen as they are in man: for animals, since they do not reflect, are susceptible neither to the grief that accompanies pain, nor to the joy that accompanies pleasure', before finally adding that humans too 'are sometimes in a state approaching that of the beasts, when they act almost on instinct alone and simply on the impressions made by the experience of the senses: and, in this state, their pleasures and their pains are very slight' (*Theodicy* §250: GP VI 266; H 281). *Pace* Leibniz, however, it is likely that the pain felt by a bear caught in a trap is no less intense than the pain felt by a human caught in a similar trap. Toothache is toothache whether in a dog or a human. Emotional pain and pleasure too need be no less intense in animals than in humans, as some recent studies have suggested.[40] Leibniz recognized that less intense pains can be felt by humans as they are by animals, but so too it seems can more intense pains be felt by both. Discriminating the intensity of pleasure and pain across species lines is neither rational nor justified: the suffering of a horse mortally wounded on the battle-field is surely far greater than that suffered by a human being whose hand gets trapped in the door of the car.

Discounting or diminishing the pleasures and pains experienced by nonhuman beings serves neither their needs nor our own. As we argued in chapter 6, the wise understand that the love of God is shown not only in the love of our fellow human beings, but in the love of all creation. The interconnected relationality and interdependence of all existing things (chapter 5) makes it impossible for us not to be affected by both the pleasurable and the painful experiences had by others, human and nonhuman alike. Others in turn cannot remain unaffected by our own pleasurable and painful experiences. Accordingly, as we amended Leibniz's own words, 'no one can be truly happy in the midst of any suffering whatsoever, irrespective of which life forms are affected'.[41]

We are now in a position to see how this works in practice, according to the theory of harmonious communication or expression outlined earlier in this chapter. In the long quotation from his 9 October 1687 letter to Arnauld cited earlier, we saw Leibniz explain that the continuity and divisibility of matter ensures that 'the smallest movement extends its effects over neighbouring bodies and consequently from neighbour to neighbour *ad infinitum*', so that our own body and indeed all bodies 'must be affected in a way by the changes of all the others'.[42] The physical states produced by these motions are reflected in each and every soul. First they are perceived directly by the soul or entelechy in whose organic body the motions occur, and then indirectly in other souls and entelechies as they in turn perceive their own bodies and the effects made upon them by the motions of the first body. It follows from this, that although we may discount as insignificant or even be completely unaware of the pain and suffering of the factory-farmed hen and the disruptions to the mechanism of its body that manifests in the loss of its feathers, these sufferings and disruptions nevertheless have a corresponding influence on our bodies and our minds. The disorganization of the hens influences how their bodies affect other bodies throughout the world, including our own, as they enter the food chain through the hens' eggs and as meat, or into the wider environment through the hens' waste products. The hens' suffering, expressed in the disorganization of its body, is also communicated to our and others' souls, expressed by us as the disruption caused in our bodies through their physical connections with the hens. Similarly, the effects of the disruption to the organic bodies of fish entailed by the ingestion of plastics that we humans have deposited in the world's oceans, ripples through the physical world and eventually rebounds on our own bodies through the food and other causal chains. Once again, too, the union of soul and body in each living creature ensures that the suffering of the fish that accompanies the damage to its body is also communicated to our minds and to the souls or entelechies of all other living creatures as they in turn express the damage to their bodies resulting from their own bodies' causal connections to the fish. Like an epidemic, the pain of any one of God's creatures is communicated to the rest.

The same point can be made in terms of the harmony and beauty of each soul as a representation of the world. When the harmony of the world is disrupted in its parts, this is reflected as a disharmony in each soul's perceptions of that world. The order implicit in the soul's perceptions as one of its perfections is not as perfect as it might be when too much of what it perceives is disordered and disharmonious.[43] Aside from the fact that any reduction of the overall beauty of the world diminishes the amount of pleasure that a rational mind might derive from the contemplation of that beauty, each mind, soul or entelechy is better served when its internal perceptual qualities are perceptions of ordered harmonious variety than it is when its perceptual representations of the world reflect, whether confusedly or distinctly, disorder and disharmony, pain and suffering in the parts of that world. When a soul's perceptions are perceptions of disorder and disharmony or perceptions of pain and suffering, then to that extent also the soul itself is disorganized and pained. Again, it appears that insofar as any one creature is harmed, all are harmed.

When these physical and psychical connections are only subconsciously registered, when the soul fails to perceive the connections distinctly, the associated lack of appetitive force that produces only confused rather than distinct perceptions also ensures that little or no action to reverse the situation is taken. However, when they are perceived distinctly, as for instance in the distinct perceptions had by the wise (chapter 6), the appetitive force that produces the distinct perceptions takes the form of a rational will or volition and results naturally in the exercise of charity or universal benevolence that seeks the perfectibility of all living things as far as possible. The wise have some distinct perception or knowledge of what is truly good. They understand the interrelatedness of all things; that their identities are bound up with each other; and that all are important in respect of the perfection of the whole. They appreciate that the good of others constitutes their own good too. Their love of God, arising from their appreciation of God's perfection, is manifested as the love of God's creation, love that is given joyfully and benevolently, motivated not by selfish concerns, but simply by the desire or will to promote the wellbeing of the other.

However, passivity in the form of primitive passive force is a feature of all created monads.[44] Consequently, all monads have some confused perceptions, for their appetitions (as modifications of primitive active force that is limited by passive force) are never strong enough to reach perceptions that are fully distinct.[45] Even the most wise and rational person does not have perceptions sufficiently distinct to perceive the whole with absolute clarity and distinctness, to perceive each one of the infinity of creatures that comprise the world with total distinctness. Each of us is limited to the extent that we perceive or express the world only from our own perspectives, from the positions of our own bodies. What we know of others is filtered through the body's sense organs and their causal connections to external bodies. Confused sensations of bodies are indispensable to this process.

Nonetheless, as self-conscious rational beings we have perceptions distinct enough for us to be able to recognize ourselves as individuals different from others and to realize that others are themselves living beings that perceive the world from perspectives that are different from our own. The essential counterpart to recognition of our own distinctive individuality is our recognition of the unique individuality of the other. The individuality or distinctiveness of the other is perceived by us distinctly when we are able to identify marks or reasons that distinguish the individual thing from ourselves and from others.[46] The most easily identifiable individuating marks are the physical qualities of the distinct individuals. For instance, bats are easily distinguished from humans by their physical appearances. However, these alone do not tell us a great deal about how bats experience the world from their perspectives. Reasoning from the perceived differences in the organic bodies of things lets us only a little way in to the viewpoints of others. Of course we can infer perfections of a creature's soul from the perceived perfections of the organic body, but it is not enough merely to grasp abstractly and detachedly by reason *that* the bat hears rather than sees the world. What is required for a more distinct perception is a sense of *how it feels* to navigate the world by sonar. This requires that distinct perceptions and reason are combined with confused sensory perceptions and imagination. It demands that we empathize with the other in such a way as to *experience*, as far as possible, the world from their perspective while at the same time keeping a firm grasp of the world as it is perceived from our own perspectives.

Leibniz himself did not coin the term 'empathy'. It is not uncommonly attributed to Lotze's nineteenth-century use of the German term *Einfühlung*, translated as 'empathy' or 'in-feeling' (from the Greek *empatheia* (passion), *en* = in + *pathos* = feeling).[47] Nevertheless, the concept of empathy is one that Leibniz advances as the act of 'putting oneself in the place of the other' (*The Other's Place (La place d'autruy)*: Grua II 699: Dascal 164). This involves attempting to perceive the world not only from one's own perspective, but also from the other's point of view, so as to include, as it were, the other's point of view within one's own. Others' bodies are the physical manifestations of each monad's point of view. Given the nature of the expressive relationship of all things through their respective perceptions of each other's bodies, it follows that the others' points of view are, in a sense, already included in each monad's perceptions. But they are perceived only insensibly or confusedly unless, or until, a conscious effort is made to perceive the nature of another distinctly by actively striving to understand and to experience as best we can the world from their perspective. When successful, it leads one to the realization of the reality of the other's experience, not in the abstract, but in as much particularity as possible. Closely related to sympathy (feeling-with) and especially to compassion (co-suffering), the empathic putting of oneself in the other's place is also a way in which one feels another's pain – and is pained by it – while remaining conscious of the fact that it is the other's pain that one is feeling, *not* one's own. This makes it possible to deeply realize in oneself the pain or suffering of the other without being debilitated by it, for one retains one's original own point of view in addition to incorporating the point of view of the other into one's own.

Successful empathy requires that we adopt an imaginative fiction: the identification of the self with the other. We use our imaginations to see the world as they do; to experience their feelings, desires and emotions *as if* they were our own, or *as if* our souls were experiencing the world from inside their bodies instead of from within our own bodies. Noa Naaman Zauderer draws attention to the importance of imagining oneself in the place of the other as a wise method for overcoming our self-interested and biased inclinations. In her words: 'Only if one actually *imagines* himself in the *concrete* place of the other, under the conditions that shape his choices, will he be able to transcend his own self-interested perspective and discover insights that would otherwise not occur to him' (Naaman Zauderer 2009, 316).[48]

Naaman Zauderer goes on to unpack the connections between the imagining of the place of the other and Leibniz's notion of justice as universal benevolence or the charity of the wise and his notion of disinterested love in which we take pleasure in the perception of the happiness of the other. When we love another, we will what we trust will make them happy. But first, we must find out in what their happiness lies. As we saw in chapter 6 and as Naaman Zauderer also points out, when we truly love God, we will what God wills. We strive to determine what God wills and then seek to implement it. Something similar is the case in empathic relations to others. The empath strives 'to transcend his own self-centred viewpoint and "as if" intrude into the peculiar perspective of another' (Naaman Zauderer 2009, 320). The technique is central to the implementation of justice as universal benevolence that Leibniz conceives as benevolence towards all humans, but which we suggested (chapter 6) can be extended to embrace benevolence to all living creatures to become an ecological justice, the charity of the wise directed towards all living beings. To love others and to act with justice towards them, whether they are human or nonhuman, we need first to enter their worlds in order to discover what is truly in their best interests and to resist the self-centredness that makes it all too easy to assume that their needs and wants are the same as our own.[49]

Leibniz advocated the putting of oneself in the other's place in his 1679(?) essay of that title.[50] There, he explores the moral and political benefits of the technique. These benefits concern relations between humans – protecting oneself against possible malign political designs of counsellors or heads of state or, in morals, giving pause to consider the possible effects of one's actions on one's neighbour or to acknowledge that 'our neighbour is not so ill-meaning or even so clear-sighted as I suppose' (Grua II 700; Dascal 164). From a practical point of view, Leibniz thinks it is prudent in political matters to assume the worst and take precautions against possible attack, but in moral matters, 'when what is at stake is harming or offending the other' (Grua II 700; Dascal 164), it is advisable to assume the best. In both cases, as Dascal reports, putting ourselves in the other's place gives us the opportunity to take into account 'the other's circumstances, needs, and goals' (Dascal 163).[51] It allows us to evaluate the essential needs of the others against those that are less important or that might be satisfied in different ways, and opens up ways that we might moderate our own demands and needs in similar fashion (Dascal 163).

Although Leibniz advocates the 'other's place' strategy only with respect to our relations with other human beings, given the harmonious communication between all living things and the translatability from visible bodily states of others (the perfections of the body) to their invisible psychical states (the perfections of the soul), it should in principle be possible to put ourselves in the place of any other living thing, human or nonhuman. Developing empathy for nonhuman beings encourages us to take their needs and interests into consideration when contemplating how we will act, and impels us to moderate or restrain our own behaviour when it might otherwise pose a threat to them.[52] Putting oneself in the place of the spider whose web we are about to destroy may lead us to let well alone. We might, as we put ourselves in the place of the spider, imagine how we would feel were our home and vegetable garden suddenly and indiscriminately destroyed. Knowing how clean and fresh we feel when our bodies are scrubbed clean of dead skin or how light and energized we feel just after our hair has been cut and tidied, we might, in empathizing with the tree, be encouraged to strip away its dead wood to allow the air to circulate more freely and stimulate new growth. This is not so much an *anthropo*morphic exercise as a *bio*morphic one. We understand that we ourselves are living beings and that, as such, the perfections of our bodies reside in the organization, motion and force of our bodies. We understand too that the perfections of our souls reside in their perceptions, appetitions and primitive force. Moreover, we feel the connection between these perfections of our bodies and those of our souls. When we put ourselves in the place of a nonhuman living thing, we recognize the commonalities between the perfections of our living bodies and the perfections of other living bodies and extrapolate imaginatively to the perfections of their souls or entelechies. These commonalities are based in the fact that we are all living things. Our humanity is only the way that this life is manifested in our own cases; life is manifested in many other ways in nonhuman beings too.

The imaginative putting of oneself in the place of the other – the seeing of the world from the other's perspective – is exactly the sort of understanding of the other that is necessary to motivate the kind of compassionate care for the other that leads us to act with their best interests at heart. Contemporary evolutionary ecologist, Marc Bekoff, regards evolutionary continuity as central to the development of compassionate attitudes and behaviour towards nonhuman animals.[53] However, in light of the discussion in this chapter, I would suggest that it is not so much evolutionary continuity that is the immediate condition underpinning our concern for the nonhuman animal, but rather the harmonious communication that operates across the Leibnizian world. Empathy is the way of making what is communicated louder and clearer. The difference between the confused expression of another's physical and psychical condition through sense perceptions of its organic body and the more distinct expression of these states that is achieved by combining reason with the imaginative adoption of the other's point of view as if it were one's own is like the difference between receiving a letter and actually opening it and reading it. We have the information within us; we just need to access it. And just as we cannot respond to the requests in the letter unless we open it and read it, so we will not act on what we implicitly and confusedly already know about the states of others until we more fully comprehend them. But when we do attain more distinct perceptions of the states of others, we will, according to the Leibnizian metaphysics, act in response, for the primitive active force that is modified as the soul's rational appetitions towards distinctly perceived true goods is simultaneously modified physically as its body's derivative active force or power to move towards that which is truly good. The power in us that allows us, by combining reason with imagination, to enter the worlds of others and to imaginatively experience the world from their points of view as well as from our own, is the same power that grounds our feelings of pleasure in the distinct perception of others' perfection and wellbeing, our will to further their interests, and the physical power that moves us to act on their behalf.

In this chapter, we have been considering the communication and empathic relations of coexisting or contemporaneous living things. In the following chapter, we turn to Leibniz's understanding of time as relational. This leads us to examine the relations that hold between past, present and future states of living and non-living things and to consider the possibility that our empathic feelings might also be directed towards beings in the future.

Notes

1 Montaigne 1987, 17–18, 23–24.
2 For a sympathetic overview of the contemporary debate regarding the waggle dance of the honeybees as language, see Crist 2004.
3 GP IV 493–498; WF 47–52.
4 GP I 424–427 and GP IV 487–490; WF 41–44.
5 I.e. the scholastic Aristotelian intentional or sensible species. See Hatfield 1998, 956–959.
6 Leibniz's explanation in the *New System* itself had been somewhat ambiguous as to whether 'communication-by-agreement' should be classed as 'communication'. There, he had explained that 'since each of these substances accurately represents the whole universe in its own way and from a particular point of view, and since its perceptions or expressions of external things occur in the soul at just the right time in virtue of its own laws ... there will be a perfect agreement between all these substances, which produces the same effect as would be observed *if they communicated with one another by means of a transmission* of species or qualities, such as most ordinary philosophers suppose' (GP IV 484; WF 18, my emphasis). He then goes on to note that this 'mutual relationship, arranged in advance in each substance in the universe ... produces *what we call their communication*' (GP IV 484; WF 18, my emphasis). But although we may call this mutual relationship 'communication', towards the end of the paper he specifically denies the 'perfect agreement' between substances constitutes 'communication' when he argues that 'this perfect agreement of so many substances *which have no communication with one another* could come only from their common cause [God]' (GP IV 486; WF 19–20, my emphasis).
7 See previous note, note 6.
8 Basnage published extracts of Leibniz's letter of 3 January 1696 in the February issue of the *Histoire des Ouvrages des Savants*. This *Second Explanation of the New System* was followed by a *Third Explanation* in the November 1696 issue, parts of which were also taken from Leibniz's January letter (WF 61).
9 Chapter 2, p. 41.
10 In Holy Communion, celebrants share in the body and blood of Christ. Members of a community share common beliefs, ways of life, aspirations and goals.
11 Kulstad 1977 remains the most helpful starting point for understanding Leibniz's notion of expression. See also Swoyer 1995 and Kulstad 2006.
12 That Leibnizian perceptions are perceptual states was first noted by Kulstad (1982, 72).
13 See also *New Essays*: A VI vi 403; RB 403.
14 As Swoyer puts it, 'a perception *expresses* the shapes and motions of the small particles that cause it, because the myriad, subliminal perceptions ("minute perceptions") that compose the perception correspond to the motions of the tiny particles that produce them – a correspondence that Leibniz likens, as in so many other discussions of expression, to that between the points on an ellipse and those on a circle' (Swoyer 1995, 92).
15 The claim is made frequently, as for instance in the correspondence with de Volder: 'any soul or entelechy whatsoever expresses both its own body, and, through it, everything else' (to de Volder, 20 June 1703: LV 266–267). See also *Monadology* §62.

16 Equally, as relational states, the soul's perceptions and the body's figures and motions (see chapter 5) accord with the definition of expression that Leibniz offered as early as 1678 in his unpublished paper, *What Is an Idea?*: 'That is said to express a thing in which there are relations [*habitudines*] which correspond to the relations of the thing expressed' and that therefore allow us to 'pass from a consideration of the relations in the expression to a knowledge of the corresponding properties of the thing expressed' (G VII 263–264; L 207). See also: 'One thing *expresses* another (in my terminology) when there exists a constant and fixed relationship between what can be said of one and of the other' (to Arnauld, 9 October 1687: GP II 112; LA 144).

17 Although isomorphism is a feature of perceptual expression, it is not necessarily present in other types of expressive relations. See Swoyer 1995, 81 for examples of non-isomorphic expression. See also Simmons 2001, esp. 67. Bolton 2011, 138–143 is a helpful discussion of different types of expressive relations in Leibniz.

18 Just as we do not perceive distinctly the individual blue and yellow particles in the green object or each individual wave that makes up the roar of the sea, nor do we 'distinctly perceive all the movements of our body, as for instance that of lymph' (to Arnauld, 9 October 1687: GP II 113; LA 144–145).

19 Rutherford et al. (2012) have demonstrated high degrees of accuracy in human qualitative assessment of the emotional states of pigs based on their observed behaviour, discovering a strong correlation of qualitative assessments with expectations following drug-induced neurophysiological manipulations of the pigs.

20 To Arnauld, 9 October 1687: GP II 114; LA 146.

21 Leibniz himself endorses proceeding in this manner in relation to rational spirits or minds. See *New Essays* 3.6.12: A VI vi 307; RB 307. See chapter 4, p. 76.

22 This is in keeping with Leibniz's definition of a clear and distinct idea as one that enables not only the object of perception to be distinguished from other objects (e.g. the perceiver and the perceived), but also, given that in a distinct perception, some of the parts are also clearly perceived, reasons for the distinction, based on their having different parts, can be stated. See chapter 6, pp. 116–117.

23 In chapter 4, we suggested that creativity is a quality that the human soul shares with God, as do nonhuman entelechies to some degree. See chapter 4, pp. 73–74.

24 Leibniz goes on to add, 'Thus there is no doubt that he who consorts much with excellent people or things becomes himself more excellent' (*On Wisdom*: GP VII 86; L 425). We may presume the converse is also the case.

25 Chapter 4, p. 76.

26 See also, *Monadology* §§42, 48.

27 That minds, souls and entelechies differ in the degrees to which they manifest their perfections hints at a hierarchical structure that could be used to imply that some creatures have greater worth or value – more perfection – than others. To some extent, this is Leibniz's view (e.g. see chapter 5, p. 107), but it should be remembered that this greater worth is calculated according to ability to assist others, and is therefore perfectly in keeping with the position developed through chapters 4 and 5 that all living beings have value both in themselves and as constituents of the whole.

28 Monads' natural indestructibility and rational soul's immortality (*Monadology* §77; *Theodicy* §90; *Principles of Nature and of Grace* §2; *On the Ultimate Origination of Things*: AG 154) also seem to resemble the absolute eternality of the divine.

29 Translation by Strickland, *Leibniz Translations.com*, <http://www.leibniz-transla tions.com/locke.htm>.

30 In fact, Locke himself declares in his Third Letter that he regards sensation as a form of thinking (Locke 2008, 495).

31 Chapter 2, p. 38.

32 Chapter 4, pp. 70–71.

33 Derivative active force gives rise to the motion of bodies, either as 'solicitations' to motion (dead forces) or as forces of an actually moving body (living forces). See

Specimen Dynamicum: GM VI 238; L 438. Garber (2009) provides detailed analysis of Leibniz's physics and traces the development of his thought from his youthful treatises on motion to his mature criticisms of Descartes' laws of the conservation of motion. Smith and Phemister (2007) explore Leibniz's account of the derivative forces of preformed organic bodies.

34 In this way, motions in organic bodies are the efficient causes that map directly to the appetitions that are the final causes in the soul.

35 See above, pp. 135–136.

36 So too, as we shall see in chapter 8, what is occurring now is influential in shaping the future.

37 Phemister 2015, 138–140.

38 Of course, not all disruptions to the normal organization of the body are imperfections nor mirrored as imperfections in the soul. For instance, blindness is often compensated for by the increased sensitivity in the other sense organs and in other parts of the brain. The soul or entelechy will have different perceptions, but there is no reason to believe the damage to the body's organs of sight are matched by sadness or any other adverse experience in the creature's soul or entelechy. So long as an amputated limb has been compensated for in the rest of the body, no pain need be felt in the soul. A creature can learn to walk just as well on three legs as on four.

39 Our knowledge of others' bodies is limited. Our sense perceptions of them are confused. Without clear and distinct perceptions of all of their infinitely many parts, we can never be certain that the mechanism of any living body is functioning properly in *all* its parts.

40 See, for instance, Masson and McCarthy 1996; Bekoff 2007.

41 Chapter 6, p. 126.

42 GP II 112; LA 144. See above, p. 136.

43 Leibniz admitted that some disharmony can have a positive effect (chapter 4, note 25 (p. 91)). In this regard, consider again the plight of the factory hens. One positive effect might lie in the raising of human consciousness to the point where such practices were considered wholly unacceptable. Questions relating to the perfection of the world and its progress towards greater perfection and issues concerning the determinism implicit in the doctrine of pre-established harmony will be addressed in chapter 8.

44 Otherwise, they would be indistinguishable from God for they, like God, would be absolutely perfect in all respects (*Theodicy* §31: GP VI 121; H 142). Without some passivity, all their perceptions would, like God's, be completely distinct (*Theodicy* §64: GP VI 137; H 158).

45 As we saw earlier in this chapter (p. 135), even the most distinct abstract thoughts contain an infinity of confused perceptions. Minds must perceive the world at least in part through their confused sensations or sensations of bodies.

46 Chapter 6, pp. 116–117. See also above, note 22 (p. 152).

47 *OED*, entry for 'empathy'.

48 Imagination, of course, is never guaranteed to capture the whole truth of the other's reality. We do not have a divine knowledge of others' experiences. Our access is inevitably prone to error for it is grounded in our confused perceptions of their bodies and our imaginative placing of ourselves in their situation. But we must try nonetheless, by imagining ourselves in their place, to determine how the other might wish to be treated, and act accordingly.

49 We may sometimes be morally obliged not to satisfy the actual desires of the other, if, for instance, their satisfaction would lead to the other's self-destruction or inflict harm on others.

50 *The Other's Place*: Grua II 699–701; Dascal 164–166.

51 Of course, if the goals are found to be malign, rather than satisfy them we will instead make moves to protect ourselves against the other.

52 As Hume recognized, it is easier to assume the perspectives of those with whom we already have much in common: 'we find, that where, beside the general resemblance of our natures, there is any peculiar similarity in our manners, or character, or country, or language, it facilitates the sympathy' (*Treatise* 2.1.11). It is harder to imagine ourselves in the places of nonhuman others. Nonetheless, as we have argued in this chapter, we do share a great deal in common with all other living creatures. We all perceive the same world and although it requires a huge leap of the imagination to think ourselves into the places of fish, plants or even into the places of cells, bacteria, atoms, quarks, neutrinos and so forth, it is not wholly impossible. For an artist's study of empathy in relation to nonhuman living things, with a particular focus on human empathy with trees, see Goto Collins 2012. For a fictional attempt to imaginatively inhabit the microbiological world, see L'Engel 1973.

53 Bekoff 2012.

8 Past, present, future

Time present and time past
Are both perhaps present in time future
And time future contained in time past.

(T. S. Eliot, 'Burnt Norton', Four Quartets)

Leibnizian time, like space, is relational, not absolute. It results from the relations that pertain to actual substances or things. Space is the 'order of co-existents'; time is the 'order of successive existents' (to Conti, November or December 1715: LC 185)[1] or the order of the successive states of existing beings. Physical space and places arise from the relations between bodies, founded ultimately upon the relational properties of the monadic substances. From this, in chapter 6, we developed an account of relational psychical space and places based upon the relational qualities of the monads, namely, their perceptions and appetitions. This opened up the possibility of describing space and particular places in terms of moral, aesthetic and spiritual values. It was suggested that psychical space as a whole, i.e. the aggregate whole composed of all particular places, is equivalent to the kingdom of final causes, while physical space comprises the parallel kingdom of efficient causes. Taken together, particular physical places make up relational physical space. In similar fashion, Leibniz conceives relational time as composed of particular 'temporal places' (Vailati 1997, 121). That is to say, periods of longer or shorter durations together comprise time as a whole, the past, present and future of the world.[2] Although Leibniz conceives time in terms of durational quantities (Leibniz's Fifth Letter to Clarke: GP VII 404; LC 75 and GP VII 415; LC 89–90), it will later prove helpful from an ecophilosophical perspective to proceed as we did for space and to focus on the moral, aesthetic and spiritual relationships of substances and things, as they connect beings in qualitative axiological ways across time.

The concepts of possible individuals in possible worlds include relational predicates that specify spatial and temporal orderings of these possible entities. Prior to God actively creating those beings that comprise the actual world, such orders are only conceptual possibilities. Even after creation, time itself, like space, considered in itself abstracted from the things that are related, is only an ideal structure.[3] However, there is also a sense in which time and space are real or actual:[4]

The parts of time or place, considered in themselves, are ideal things; and therefore they perfectly resemble one another like two abstract units. But it is not so with two concrete ones, or with two real times, or two spaces filled up, that is, truly actual.

(Leibniz's Fifth Letter to Clarke: GP VII 395; LC 63)

Actual time is more than the mere conceptual ordering of things. Relational places and times come into being at the moment of the creation of things that stand in actual relation to each other. Actual or 'real' time is the ordered flux in which there is movement or unfolding of existing beings as their present becomes past and their future becomes the present. The relational properties internal to living substances are the foundations upon which actual times are built. From the relations of existing things, both their relations to themselves and their relations to others, places and times are constructed and particularized. Longer durations of time have more successive and 'like states interposed' than do shorter durations (Leibniz's Fifth Letter to Clarke: GP VII 387: LC 89–90). Leibniz here offers no further details, but the suggestion is made in response to Clarke's objection that mere succession of things does not guarantee that they succeed each other at the same rate. For Clarke, time must be an absolute quantity within which things proceed successively, some faster, some slower than others (Arthur 2014, 164). Leibniz appears to offer the quantity of 'like states interposed' as a way of allowing relational times and durations to be consistently measurable quantities of greater or lesser duration, analogous to the measurement of clock time in hours, minutes and seconds.[5] Although Leibniz does not state so explicitly, it is not unreasonable to suppose that different time periods are also qualitatively distinct. Indeed, they must be so, since the 'states interposed' are founded upon the relational properties of qualitatively distinct individual things. In this sense, the durational parts of time differ insofar as they arise from relations between actual beings whose constantly changing perceptual experiences mirror the flux of events from each individual's own perspectival point of view. So, for instance, even though they may be quantitatively of the same durational length, the times we pass in the company of friends are qualitatively different from the times that we spend with colleagues at work, each being founded upon the internal relational qualities of different individuals.

If, as Leibniz believes, actual time arises only when actual existents and their states stand in successive relations to each other, the question of when creation happened becomes nonsensical and the related question as to whether it could have occurred earlier or later than it did must be answered in the negative. When these questions arise in Leibniz's correspondence with Clarke, Leibniz points out that strictly speaking, questions as to the timing of the creation only properly arise under a theory of absolute time, for only as absolute is time itself regarded as 'distinct from things existing in time' (Leibniz's Third Letter to Clarke: GP VII 364; LC 26–27). Under a theory of relational time, there is no real time before the instant of creation. Time only comes into being at the moment of creation. Creation could not have occurred earlier because there was no time at all before creation.

However both Leibniz and Clarke are aware of the threat this posed to the omnipotence of God for it appears to make it impossible for God to determine 'when' to create the universe. On his side, Leibniz proffers two solutions, both of which he ultimately rejects. First, we might suppose that God could have created more substances than he did and that relations between these might have constituted a time prior to the actual beginning of the world. This is rejected on the ground that God, aiming to maximize variety, would have created such substances if it had been possible to do so (Leibniz's Fifth Letter to Clarke: GP VII 404–405; LC 75–76). Second, Leibniz proposes that God could have created, in the first instance, only 'immaterial creatures' without also creating the bodies to which they are united. The relations between these disembodied souls would be sufficient to start the temporal ball rolling. If, sometime later, God decided also to create matter or bodies to accompany these disembodied souls, God would have a choice as to when to create these bodies. Of course, as Leibniz confidently asserts, there are no disembodied souls and so this option too comes to naught (Leibniz's Fifth Letter to Clarke: GP VII 406; LC 77). In effect, by deciding that souls and their bodies should never be separated, God relinquished any choice he might have had to create matter earlier or later.

For our purposes, the most interesting feature of Leibniz's second proposal is its implicit acknowledgement that time can arise not only from external relations between bodies, as required in his account of physical space, but also from relations between monads' perceptions.[6] The internal relations that hold between any single monad's perceptions provide sufficient ground for time to come into being as the monad's law of the series of perceptions unfolds as the series of its lived experiences. Even if God had created only one solitary, disembodied soul or monad, the relations between its own internal perceptions would suffice to constitute time itself.[7] Minimally, all that is needed is that there are diverse psychical or mental phenomena occurring in an ordered succession in a single mind or entelechy. Commentators refer to time understood in this sense as 'intrinsic monadic time' (Arthur 2014, 161) or as 'intramonadic' (Nita 2008, 16).

Intrinsic or intramonadic time arises when the series of perceptions and appetitions that constitutes an individual's life begins to unfold, with one perception following another in the order specified by the individual's unique 'law of the series'. We know that Leibniz held that every perception is a perception of the whole universe, of all the things and all the events in the entire history of the world. Past, present and future perceptions cannot, therefore, be distinguished one from another by their content, for the content of each individual state is the same from one moment to the next. Instead, the different states must be distinguished by the ways in which this content is perceived, taking into consideration both their position in the series and their relative degrees of confusion and distinctness.[8] As I sit here 'now', I see my fingers on the keyboard, I feel my body supported by the chair, I sense perceive the tree outside the window, blowing in the wind, I watch its movement and the changes in its surroundings, but when I try to recall what the tree was like even just a moment ago, I realize that I am no longer sense perceiving it as it was. I can remember it vaguely, but the vivid sensation of its green leaves and brown trunk, the vision of the crow hovering just above it – these determinate sensations have passed. As I look now, I see the crow return and I sense again the brown and green of the tree, but these are new experiences, not exactly the same as the earlier ones. The former sensations are past and will not be experienced so vividly again. I am left with only vague memories of how the tree was in the past. My present moment places me in a particular place and time, replete with confused sensations and distinct thoughts. My memories of the past are less distinct even than my current confused sensations – in Humean terminology, the past is no longer an impression, but only held in the mind as an idea.[9] As a representation of the whole universe, my present perceptions also contain a representation of everything that has yet to be. The present is already 'pregnant' (*gros*) with the future. Again, however, like memories, future events are also perceived only minutely in the present. Only when my perceptions of them become more distinct will what is now future be experienced by me as my present. In this way, within a monad's series of perceptions, the ordered sequential unfolding of a monad's experiences can be described in terms of the relative degrees of confusedness and distinctness of particular perceptions of the same event or thing, such that, within the series of perceptions that constitute this life, what is in the future is that which will be, but has not yet been, perceived more distinctly; what is past is that which has been perceived more distinctly, but will not be perceived so distinctly again; and what is present is that which is, relatively speaking, perceived distinctly and will not be perceived so distinctly again.[10]

The representative nature of each monad by which each expresses the state of the whole universe at each and every moment ensures that when God creates the myriad of monads that comprise the real world, the unfolding in parallel of their individual series of perceptions and appetitions constitutes what we may call 'extramonadic time': actual time that is a 'unique succession of all the sets of simultaneous states of the other monads in the same universe' (Arthur 2014, 162).[11] As the essences of the monads progress from one perceptual state to the next, each successive state of one monad corresponds to the states of the others, either as past, present or future. Leibniz's nonabsolute relational time is therefore not only internal to each monad, but is also an ordering of one individual's changing, psychical states relative to those of all the others. And since every monadic soul or entelechy is attached to an organic body whose motions and resistances correspond exactly to the perceptions and appetitions of its dominant entelechy, we may add to our understanding of time as both intra- and extramonadic, a third: time as 'extracorporeal', time as the succession of changing relations between substances' organic bodies. Thus, through the relations and mapping of one monad's sequential perceptions to those of the others and the corresponding relations between sequential states of their organic bodies, the flow of extramonadic and extracorporeal (or public) time comes into existence.

Souls' perceptual states mirror the states of their organic bodies, recording moment to moment the changes wrought on their bodies through their interactions with external things.[12] Strictly speaking, therefore, because of the time it takes for light or sound to travel from an object to our eyes, our indirect sense perceptions of external things are perceptions of things not as they are now, but of things as they were in the past. The time that elapses while light travels from a distant star to our eyes means that our present perception of a star is a representation of a past state of the star itself. Indeed, the star itself may have long since ceased to be by the time that we see it. Our perceptions of the states of things in our immediate environment are also subject to time delay. As I watch a worm burrowing into the soil, there is a temporal lapse between the worm's sense of its own movement and my perception of its movement. Clearly, the finer detail of Leibniz's theory of time is in need of some modification, the addition of a layer of complexity that Leibniz himself did not provide.[13] Nevertheless, Leibniz's general intuition stands firm: time is the order of things as they exist in succession; space is the order of coexisting things. As the ordering of past, present and future, time is the order of things that do not coexist, i.e. the order of things insofar as they do not coexist in the present.

The temporal order of things is an unfolding order of final and efficient causes as creatures' appetitions take them from one perception to the next and the physical forces of their organic bodies effect changes in their relative spatial positions. New perceptions and physical states, as we saw in chapter 7, [14] are never completely severed from the states that preceded them, nor from those that will follow. As causes, past events are retained in the new present moment through the effects that they have produced. The past is always with us, for the 'entire effect must always express its [entire] cause' (to Sophie, 6 February 1706: GP VII 570; LS 350).[15] Each present moment contains traces of everything that has preceded, for all of the past has been instrumental in the formation of this present moment.[16] Moreover, since every future effect, when it in turn becomes a present moment, will also express its entire cause, including this current present moment, each entire present effect is also a future cause. The future is already contained in the present insofar as the present effect–cause points towards the infinitely many future effects that it will produce. '[T]he lineaments of the future are formed in advance' in each and every soul (to Arnauld, 30 April 1687: GP II 98; LA 123). Thus does every present effect–cause face both backwards and forwards. In facing forwards, a creature's appetitions strive towards what it perceives, rightly or wrongly, as the best. Although sometimes maintaining the status quo is the focus of a creature's desire, at other times, the desire is to bring about change for the better, as when a person strives to attain a more comfortable lifestyle or a wolf seeks food to assuage its hunger. In such cases, there is an implicit assumption that the future could be better than the present or the past. In the following section, we address Leibniz's position on the issue of progress, of individuals and of the world as a whole.

Perfection and progress

Together, Leibniz's assertions (i) that each physical and psychical state of every living creature in the universe is a representation or expressive mirror of the whole universe, past, present and future, and (ii) that God created the best of all possible worlds, entail that every living thing at every moment is, both physically and psychically, a representation of the most perfect world possible. It is tempting to suppose that this also entails that each soul or entelechy, insofar as it is a representation of a perfect whole, is also perfect and that every organic body, as the physical correlate of its dominant soul or entelechy, is similarly perfect. However, this supposition is not justified, for although every finite substance represents the best or most perfect world, none represents this world perfectly. Only God has a fully distinct and adequate perception of the whole; only God is perfect. Finite substances represent God's creation as best they can, but they are always limited by their own imperfections. The limited degrees of perfection of a soul, manifested in terms of the forcefulness of its appetitions and the distinctness of its perceptions, and the corresponding perfection of its body, calculated in terms of its organizational complexity, efficient functioning and motive force,[17] serve to differentiate one substance from another. Moreover, even within one substance itself, its own passing states are differentiated one from another according to the same criteria. No one substance retains exactly the same degree of perfection from one moment to the next, sometimes increasing, at other times diminishing.

Is it possible that some individual substances might increase in perfection indefinitely? At first glance, the supposition seems implausible. After all, whatever is born eventually dies, its body contracts and its perceptions once again collapse into insensibility or become even more insensible, as in the case of one in the 'abyss' whose perceptions are already insensible (*On the Ultimate Origination of Things*: GP VII 308; AG 155).[18] Leibniz himself, however, does not dismiss the idea of indefinite progress of individuals. He explores the possibility in a short piece that asks, under the title by which it is known, the related question, *Whether the World Increases in Perfection*. In this essay, he maintains that a substance's increased perfection is not necessarily incompatible with there being periods in which it appears to regress, for the substance may still be regarded as having increased in perfection overall provided 'it can be recognized to have increased more than it has decreased' (Grua I 95; ST 196). Later in the same piece, he asks, somewhat elliptically, whether we might think of souls as increasing in perfection over time because each successive perception includes more of the past than did the immediately preceding perception:

> ... even if we do not remember distinctly, nevertheless, the whole that we now perceive consists of parts, into which all preceding actions enter. So should souls always progress over time towards more clearly expressed thoughts?
>
> (*Whether the World Increases in Perfection*: Grua I 95; ST 197)

Leibniz appears to be toying with the idea that souls might be continually increasing in perfection *because* their past experiences are retained within their present states. Of course, since all past and future states of a substance are already represented in the present, past events, even before they were more distinctly perceived as present events, were already contained in the substance whose timeline contains them. However, before they actually occurred, the timeline contained them only as minutely perceived future events. On becoming more distinctly perceived as present events, they become causes rather than would-be future effects, and their influence continues, even though perceptions of them recede once more into insensibility as the present itself becomes past. We can understand this in terms of learning from the past. As we ourselves make the transition from childhood into adulthood, we accumulate experiences, adjusting and refining our assumptions and opinions in their light and hopefully gaining increased understanding of ourselves and of others along the way. While there are periods, such as during sleep, when our perceptions descend into obscurity and insensibility, when we wake again, we retain and build upon our past experiences. For the most part, the overall trend is upwards. Human rational souls become increasingly more perfect as their knowledge increases with experience. In contrast, '[t]he greatest being [i.e. God who perceives all with the utmost clarity and distinctness] does not increase in perfection, because it is outside time and change and it includes the present and future equally' (Grua I 95; ST 197); that is to say, rather than the past and future being perceived with less clarity and distinctness than the present, as in finite beings, in God, all are perceived with the same – and highest – degree of distinctness.

At the other end of human life, however, in the descent towards death, perceptions decline in distinctness, memories fade and thought processes slow down. Nevertheless, Leibniz does not consider this as a reason to reject the overall progress to greater perfection in the soul for despite the lack of distinct memories, it is still the case that present experiences build upon past experiences, and that 'all preceding actions' enter into each successive experience. Moreover, he insists that even on death, rational souls or minds do not lose their identities or memories. They retain their personalities and distinct perceptions:

> All souls keep their substance and are imperishable, even those of beasts; but only rational souls still keep their personality, that is, the reflective knowledge of what they themselves are, or consciousness.
> (To Caroline, n.d.: Klopp XI 58; LC 192)[19]

From an ecological perspective, we saw no good reason to retain Leibniz's sharp distinction between rational souls and those of other living creatures.[20] No doubt motivated in large part by the theological orthodoxy of his own age, the distinction created a point of tension in Leibniz's philosophical position, for it sits uneasily alongside his commitment to the Principle of Continuity. And indeed, with respect to both the indestructibility and the perfectibility of substances, Leibniz's remarks apply equally well to all living beings, rational and nonrational alike. All living substances are indivisible and therefore exempt from the natural destruction to which nonliving aggregate bodies are subjected when their parts are broken up.[21] On perfectibility, it is true that Leibniz does not grant nonhuman animals the self-conscious awareness that allows minds to consider themselves as perceivers, that is, as 'I's abstracted from the things that they perceive, to engage in the rational deliberative thinking, and to acquire knowledge of necessary truths.[22] Nevertheless, he does believe that some animals have conscious memories of the past and are able to engage in empirical reasoning through which they learn from their past experiences and adapt their behaviour accordingly.[23] Even the 'bare monads' that have only insensible perceptions retain the past in their present experiences, for in their cases too, the 'entire effect [the creature's current psychical and physical state] must always express its [entire] cause'. They too may be said to progress towards greater perfection insofar as traces of the past are discernible[24] in their present (and future) states.

Strickland (2006, 120) questions whether the developmental progression I have outlined above should be read as an increase in a substance's degree of perfection. Progress, he suggests, may in this context refer simply to the development or unfolding of a substance, understood as change, and the overall balance of these changes may be value-neutral, for although some changes result in an increase in the distinctness of a being's perceptions and hence also its degree of perfection, other changes involve a descent once again into confusion, as happens, for instance on death. However, on the proposal advocated in this chapter, the descent into confusion does not erase all of the creature's previous perceptions. On this view, each creature retains the relatively distinct perceptions they enjoyed when alive, at least insofar as they continue to influence the creature's post-death states. Its lived experiences remain within the causal nexus into its post-death state. Hence, there is a sense in which what Leibniz says of rational minds – that they retain their own memories and personality even after death – can be extended to all living beings, for although they do not remain conscious, each retains their individual identity and a semblance of memory after death. Even in death, the law of the series of their perceptions and appetitions continues to unfold, each present state building on the past as it moves into the future. In this sense, then, each creature might be said to be continually progressing – or maturing, as we shall suggest below – as its essence unfolds, for since the influence of the relatively distinct perceptions enjoyed during its life continues on into death, its overall degree of perfection can never descend lower than it was when the creature was born and its more distinct perceptions began to emerge. It has not merely changed, but rather has made real progress that is not negated in death.[25]

For all this, the jury is still out on the question whether Leibniz himself held that *all* individuals are in the process of becoming more perfect (or at least that all are in principle perfectible) or that only *some* individuals are actually becoming (or capable of becoming) more perfect. Leibniz himself appears not to have had a settled opinion on the matter, nor on the related question as to whether the perfection of the world as a whole – taken as the sum of the perfection of its parts – increases over time.[26] Strickland (2006, 115–141) tackles the complex interpretive issues involved, documenting the possible stages of Leibniz's thought from rejection, through acceptance, to eventual agnosticism. I do not intend to enter into this hermeneutical and historical debate here. Instead, I shall develop an account that draws on Leibniz's texts and arguments from various periods in an attempt to combine, in a manner of which Leibniz may or may not himself have approved, his seemingly opposed opinions: (i) that the world and its individuals do not increase in perfection from moment to moment; and (ii) that they do increase in perfection as time goes on.

In the mid-1690s, Leibniz seems to have inclined towards that opinion that the world *does not* increase in perfection over time:

> The question is whether the whole world increases or decreases in perfection, or whether in fact it always preserves the same perfection, as I rather think ...
> (*Whether the World Increases in Perfection*: Grua I 95; ST 196)

Although Leibniz takes the view in this essay that some individual creatures might increase in perfection over time, he maintains that the total sum perfection of the world can still remain the same at every moment, so long as whenever some substances increase in perfection, others decrease to a corresponding degree:

> If the perfection of the world remains the same, some substances cannot continually increase in perfection without others continually decreasing in perfection.
> (*Whether the World Increases in Perfection*: Grua I 95: ST 196)

However, in a note penned around 1689–90, Leibniz had taken the opposite view. 'All things considered', he stated then, 'I believe that the world continually increases in perfection' (*On the Continuous Increase in the Perfection of the World*: A VI iv 1642; ST 195–196). He offers three reasons why this must be so. If the world did not continually increase in perfection, then (i) there would be no final cause;[27] (ii) God would not rejoice 'in the continuous advance of his plans; and (iii) happiness, which 'requires perpetual progress to new pleasures and perfections' would be impossible (A VI iv 1642; ST 195–196). In these reasons, we find once again the underlying thought that progress in perfection involves the accumulation of present experiences from past experiences. Even though everything is already known to an omniscient God, the actual execution of God's plans, i.e. the progression of the created world towards an ever more perfect state, occurs over time, with the present building upon the past and moving forwards into the future. The unfolding of God's creation is the process through which the world gradually approaches fully realized perfection. Consistent with our earlier understanding of the increasing perfection of individual living things as building upon past experiences, Leibniz declares that the universe itself,

> is similar to a plant or animal, in that it tends towards maturity. But this is the difference, that it never comes to the greatest degree of maturity, and also that it never goes back or falls into decline.
> (A VI iv 1642; ST 196)

In *Whether the World Increases in Perfection*, Leibniz suggests (1) that the world retains the same degree of perfection at every moment, while in *On the Continuous Increase in the Perfection of the World*, he maintains the seemingly diametrically opposite view (2) that it continually increases in perfection. Despite appearances, however, the claims are not incompatible provided they are understood as appealing to different conceptions of perfection. When perfection is conceived in terms of ordered variety, the world can be considered as having the same degree of perfection at all times. The number or variety of substances remains constant and even though the relations between them are constantly changing, the changes are governed by the laws of final and efficient causes that ensure that harmonious order is maintained between souls or entelechies, between bodies, and between souls and their bodies. Perfection understood in terms of the greatest possible order and variety is consistent with the actual relations between the various constituents of the world changing from one moment to the next. When, on the other hand, perfection is conceived qualitatively in terms of moral goodness or worth, it becomes possible to conceive the perfection of the world as increasing over time. If the moral sensibilities and properties of significant numbers of individual substances are honed and matured, as for instance when greater numbers of rational beings serve others in a spirit of universal benevolence, then even though the overall quantity of substances remains constant and even though these substances continue to exist in ordered relationships, considered in moral terms the actual degree of perfection of the world itself might be regarded as having increased. In this way, the perfection of the world can be conceived as constant in one sense (ordered or harmonious variety) but as increasing in another (moral goodness).

In the *Theodicy*, Leibniz admits that he can find no reason 'why a thing cannot change its kind in relation to good or evil, without changing its degree' (*Theodicy* §202: GP VI 237; H 253). The situation is not unlike the way in which, irrespective of whether one's pleasure is sought through music or the visual arts, the degree of enjoyment obtained from each can be the same. Just as a shift from evil to goodness can be accomplished without alteration to the overall variety and order of the world, so too,

> [i]n the transition from enjoyment of music to enjoyment of painting, or *vice versa* from the pleasure of the eyes to that of the ears, the degree of enjoyment may remain the same, the latter gaining no advantage over the former save that of novelty.
>
> (*Theodicy* §202: GP VI 237; H 253)

By way of further illustration, Leibniz points to the mathematical squaring of the circle (or conversely, the circularizing of the square) as an instance in which the ordering of parts can be changed, such that what was previously a circle is now a square, but the parts remain ordered, albeit now as a square instead of a circle. Changes to the actual ordering of the parts do not entail loss of order per se among the parts: 'there will always be an order among them, and that the best order possible' (*Theodicy* §202: GP VI vi 237; H 253). Similarly, in the universe as a whole, changes to the actual relationships that hold between substances do not necessarily lead to loss of either order or variety: each substance continues to monitor at every passing moment the past, present and future states of all the others. Meanwhile, however, increases in the distinctness of substances' perceptions not only change their relationships to others, but also increase their own perfection, understood as moral perfection. As numbers grow, these increased perfections can also be considered as increasing the moral perfection of the universe itself.

As we discovered in chapter 5, differences in the relative degrees of confusion and distinctness of individuals' perceptions serve to differentiate them from one another and establish the unique points of view or positions from which each one perceives the world.[28] In chapters 6 and 7,[29] we explored the role of distinct perceptions in fostering feelings of joy, happiness, love and universal benevolence and we saw how the true perfection of rational and human beings lies in their virtue and piety, grounded in the distinct perceptions and knowledge of the beauty and goodness of God and creation. Few have yet attained the distinctness of perception of those who are truly wise, who derive pleasure from contemplating the beauty and perfection of God's creation, who love God as its source and who realize that loving God and loving that which God created are one and the same. But even the most wise amongst us frequently fail to live up to the demands that charity and justice make upon them. There is a long way to go before universal benevolence (the charity of the wise) is routinely granted to all rational beings in the kingdom of grace, and even further to go before every rational being extends this charity to all other living beings.[30] In this sense, the universe is not yet as perfect as it might be.

Not yet perfected, but the trend is upwards. Earlier in this chapter,[31] we took from a brief remark in Leibniz's *Whether the World Increases in Perfection* the idea that substances are gradually becoming more perfect as their essences unfold and their past experiences inform their present states. If this is indeed the case, then the universe itself, the sum of its constituents, is also moving inexorably 'towards maturity' (*On the Continuous Increase in the Perfection of the World*: A VI iv 1642; ST 196) or, as Leibniz states in *On the Ultimate Origination of Things*, 'progress never comes to an end', for while many 'substances have already attained great perfection', there are infinitely many more 'asleep in the abyss of things, yet to be roused and yet to be advanced to greater and better things, advanced, in a word, to greater cultivation [*ad meliorem cultum*]' (GP VII 308; AG 155). The greater cultivation or perfectibility of human and other rational beings consists in their moral development, in the development of what we, along with Catherine Wilson (Wilson 1989, 291),[32] may regard as the development of a civilized 'culture' characterized by peace, harmony, love and goodwill towards all.[33]

We have described progress in terms of the unfolding of creatures' essences and the maturing of the world as a whole. For Leibniz, such progress is certain. Post creation, the unfolding of the world is destined to follow the path already specified in laws of the series of souls' perceptions and the preformation of their organic bodies. For an ecological philosophy, the spectre of such an already predetermined future is a stumbling block that needs to be addressed. We shall therefore take this up in the next section, before turning to the question, in terms of values, of our relations to past and future beings.

Determinism and future certainties

It is Leibniz's considered opinion that events in the world, once set in motion, are certain to come to pass. He insists that everything that is now happening and everything that has yet to occur was already contained in the seeds of things right from the very beginning. The doctrine of the preformation of organic bodies or seeds[34] goes hand in hand with a mechanistic causal determinism, such that with complete knowledge of the initial conditions of preformed bodies and an understanding of the physical laws of nature, it would be possible to predict with certainty the entire future course of the world. Correspondingly, Leibniz's early metaphysical doctrine of the complete concepts of individual substances[35] dictates that everything that has, is or will be true of any one substance is already contained in its complete notion.[36] Individuals' complete concepts and the preformation of their organic bodies ensure that, both metaphysically and physically, the future of this world is certain. Having decided to create this world rather than any other of the infinitely many possible worlds he could have created, God merely set in train a sequence of events that he has already foreseen in the best possible world.

In the face of divine foreknowledge of the future of our world, we might convince ourselves that we are given licence to act with impunity, in the sure and certain knowledge that however we act, this is the best world that could ever possibly have been. We do not need to consider what might be for the best, for it is guaranteed that whatever we do, it is part of the best possible world and will have the best possible outcome. If this is the best possible world, then we can act as we please. But such a response would be naive. It matters greatly how we act, for under Leibnizian determinism, the future is certain not in spite of past and present events, but precisely *because* of them. Our deliberations as to the best courses of action, and the decisions we make in light of these, are governed by the laws of final causality. Meanwhile, the motions of our bodies through which our decisions are implemented are governed by the laws of efficient causation. Together, these laws ensure that were the choices we make and the actions we take other than they are, the outcomes would also be different and the world itself would be some other, less than best world. Hence, this world would not be the best possible world if it were not the world in which we act freely and deliberately in precisely the ways that we do. It is therefore incumbent upon us to choose our future courses of action as wisely as our capacities allow.

However, Leibniz is not quite out of the woods yet. God already knows how we will deliberate and what choices we will make. I may believe that I choose freely a certain course of action, based on distinct perceptions and rational appetite, but the fact that I will at a particular point in time perceive one course of action as better than another is already known with certainty to God. These things are already written into the law that constitutes my essence or nature. But if all the choices and decisions I will ever make have been in my essence or complete concept from the very beginning, is there any point to debating how I *should* or *ought* to act in the present? Indeed, whether I debate it or not will also be contained in my essence. All that we will choose, we will choose; all that we will do, we will do. Once God has created the world, everything that happens in it is certain to unfold in just the way that it does and not otherwise.

The problem of how to reconcile freedom and necessity is not peculiar to the Leibnizian system. Spinoza's necessitarianism encounters the same difficulty even more acutely. In Spinoza's system, the world that exists is the only world that is possible. All that occurs in that world follows from the absolute necessity of God's nature. Nothing can happen in any other way than the way that it does.[37] For Leibniz, the problem is not so severe. Leibniz's God stands apart from the world he creates – a world comprising a plurality of substances that God chose from the infinitely many other possible worlds that he might have created instead. Leibniz's God surveyed all possible worlds and then chose to create the one that was the best (*Theodicy* §§414–416: GP VI 362–364; H 370–373). Things would have been different had God chosen to create a different world.

Thus, unlike Spinoza, for whom all necessity is absolute necessity, allowing no other possibilities, Leibniz admits two types of necessity. First there is the absolute necessity of the kind typical of the truths of mathematics. Absolutely necessary truths must be true in every possible world. Their contraries 'imply contradiction' and are therefore impossible. The properties of a circle follow from the very definition of a circle, the denial of which would contravene what it means to be a circle at all. Truths about matters of fact, on the other hand, are only hypothetically necessary and need not be true in every possible world. The Leibniz who exists in this world spent some years in Paris and this fact is included in his complete concept. However, while the concept was still incomplete, that is, while the co-construction of individuals' complete concepts was ongoing,[38] we may suppose the choice of whether to go or not to go to Paris was entirely open. At the point of choice, we can imagine the (as yet incomplete) concept of Leibniz splitting in two, or rather forking out along two different pathways on the way towards more complete specification: one eventually being completed as the complete concept of the Leibniz who chose to go to Paris and who is part of the best possible world, his complete concept co-constructed in relation to the concepts of the individuals he would meet there; another eventually being completed as the concept of a 'Leibniz' who chose instead to remain in Mainz, whose complete concept would be co-constructed in relation to the concepts of the individuals there, and who would ultimately be part of a different possible world.

That our choices were made in the realm of possibilities, prior to the creation of the world, does not negate their post-creation status as contingently free.

[E]ven though it is certain that God always chooses the best, this does not prevent something less perfect from being and remaining possible in itself, even though it will not happen, since it is not its impossibility but its imperfection which causes it to be rejected. And nothing is necessary whose contrary is possible.

(*Discourse on Metaphysics* §13: GP IV 438; AG 46)[39]

God, being moved by his supreme reason to choose, among many series of things or worlds possible, that, in which free creatures should take such or such resolutions, though not without his concourse; has thereby rendered every event certain and determined once for all; without derogating thereby from the liberty of those creatures: that simple decree of choice, not at all changing, but only actualizing their free natures, which he saw in his ideas.

(Leibniz's Fifth Letter to Clarke: GP VII 390; LC 56)

In the realm of possibilities, God did not compel our possible selves to choose as we do. We determined ourselves to certain courses of action during the process in which our identities were being formed. The choices that we are certain to make once the world has been created have already been freely made *by us* in the realm of possibilities. That God then chose to create such a being knowing what it would freely choose does not take anything away from the freedom of the choice. They are still our own *self*-determined choices, freely made by ourselves, not by God. God merely chose to create the world to which we belong, knowing in advance the free choices we would then go on to make. Once created, our self-determined freedom is enacted through our spontaneous inner force (our primitive active force) that grounds our distinct perceptions and rational appetites aimed at what we think will be for the best. The greater our inner force, the more distinct our perceptions of the good, and the stronger our rational appetite or volition, then the more are we free.[40]

Axiological dimensions of time

We turn now to questions of value. Following the pattern established in the discussion of space in chapter 6, we here consider time and its temporal durations not as quantitative measurable units such as centuries, years, weeks, days, hours, minutes, seconds and so forth, but as periods of time qualitatively assessable in terms such as love and justice and the flourishing and wellbeing of the things from whose unfolding relational properties time emerges.

Temporal value relations point in two directions, backwards and forwards. Our relations to the past are to the preceding states of living things and to the aggregate bodies, living and nonliving, composed of them. Together, they comprise the complete set of final and efficient causes that have led to the present condition of the world, which state we represent or express in our perceptions. A thoroughgoing causal determinism means that everything that has happened in the past influences the present. The present moment would not be as it is now had the past been in any way different. Every previous appetitive drive of each and every entelechy has led up to the present moment and influences, but does not absolutely necessitate, the direction of monads' current appetitive drives towards the future.

When we turn our attention to the countless atrocities and other offences or injustices humans inflict upon their fellow humans and other living beings, we may well wish the past had been different, that our leaders had made different decisions or that unfavourable but unforeseen consequences had not happened. We may regret our own misdemeanours and wish we ourselves had not acted in the ways we did. Had human beings pursued different values and lifestyles, we might not now be facing anthropogenic climate change, there might be fewer or even no wars raging in regions across the globe, fewer nonhuman species in decline, and so forth.

From the standpoint of a Leibnizian metaphysics, such wishful 'what-if' thinking is futile and misguided. The past, after all, cannot be undone. Focusing on 'what might have been' can encourage feelings of regret, the holding of unforgiving grudges, and the seeking of revenge for perceived wrongdoing. It can be demoralizing, destructively depressive and potentially dangerous. However, there is a more positive attitude we can adopt towards the past. We can view the past as an instructive resource: we can use our knowledge of history to inform our understanding of the present and to help us to learn from (and to avoid repeating) past mistakes. Focusing on the past in this way can be an uplifting, hopeful and encouraging strategy to employ as we move into the future.

Acknowledging that even the most horrific wrongs and evils in the world are dissonances that serve a purpose within a whole that is the best possible[41] might lead to the consolatory but ultimately stultifying thought that no matter how horrific this world may be, all the other worlds would have been worse. However, another more positive and affirmative response is also afforded by the Leibnizian philosophy. Instead of passive acceptance, we can choose to actively seek out the good that is assuredly there in the best possible world, appreciating its ordered and varied perfection where we find it. Then, in the knowledge that its moral perfection is yet to be fully realized and spurred on by the energy and hope provided by the good we already discern, we may optimistically take positive steps to help bring about a future that improves upon the past.[42]

All the same, François-Marie Arouet Voltaire (1694–1778) pinpointed the inherent danger of Leibnizian optimism: that it threatens to belittle the very real and near intolerable suffering that is a daily occurrence in so many parts of our world. Composing *Candide* in the aftermath of the catastrophic Lisbon earthquake and tsunami in 1755 and drawing on his own life experiences, Voltaire places Leibniz's notion of the best possible world in the spotlight through the ridiculous figure of Dr Pangloss and his student, Candide.[43] The main protagonists, who include Candide's youthful sweetheart, the Baroness's beautiful daughter, Cunegonde, suffer a series of tragedies that range from whippings, conscription, syphilis, disfigurement, dismemberment, shipwreck and the Lisbon earthquake to hanging, rape, prostitution, slavery and theft.

Plus ça change ..., we may declare as news bulletins bring us regular reports of war and its atrocities, social deprivations, natural disasters and global epidemics. Voltaire's questioning of Leibniz's assertion that this is the best possible world is as relevant today as it was in the middle of the eighteenth century.

Nevertheless, there is actually remarkable convergence between Leibniz and Voltaire on specific points, especially with respect to the question of how progress may be made and maintained. Although Voltaire cast scorn on Leibniz's conception of this world as the best of all possible worlds, he did, as one translator points out, accept the ordered nature of the world and he promoted the idea of progress, implicitly in *Candide* and more explicitly in other writings (Voltaire 2009 [1759], 33). We have seen the prominence of these ideas in Leibniz's thought, but of even greater interest is that Voltaire proposed that progress could only be achieved through education and reason, and advanced, as Palmer notes, through the written word (Voltaire 2009 [1759], 34), a view that is reminiscent of Leibniz's conception of progress in terms of increased distinctness of substances' perceptions of the world as their present and future perceptions retain and build upon past experience, and his vision of the perfecting of human beings as the development of wisdom that combines distinct perception with rational appetite or will for universal good.

Progress demands that we raise our awareness of the past, the good and the bad, so that what was proven to work well might be repeated and what did not, avoided or redressed. When past horrors are acknowledged by being distinctly perceived in the present, lessons can be learnt and injustices rectified. Palmer details Voltaire's own efforts to publicize the torture and hanging in 1762 of Jean Calas, wrongly accused of killing his son. In fact the son had committed suicide, but the family had tried to ensure that the son would be buried in consecrated ground by making his suicide appear as murder. In publicizing the case, Voltaire was simply bringing the truth to light. In so doing, he opened the way to the restoration of justice. Jean Calas could not be brought back to life, but Voltaire's efforts helped ensure that Calas' name was cleared and the family compensated financially by the Crown (Voltaire 2009 [1759], 33–34). This is just one example of how past wrongs may be rectified and healthier relations with the past established. In more recent times, the trading of prosecution rights for full disclosure and transparency by the South African Truth and Reconciliation Commission may be seen as an attempt to employ distinct perception of past wrongs as an enabling tool to restore unity to a country torn apart by the apartheid regime.[44] Developing healthy relations with the past in the present is crucial if the future is to be better than the past, for the 'lineaments' of the future are already in the present. Making peace in the present with the past, therefore, would seem to be the first step towards a better future.

It is a feature of the Leibnizian account of time that we perceive current events more distinctly than we do those in the past or in the future.[45] Consequently, because of the relation between distinct perceptions and activity and between confused perceptions and inactivity, it is often (though not inevitably) the case that, as Leibniz perceptively pointed out in his *New Essays*, our appetitions towards a distinctly perceived good in the present or immediate future are stronger than our appetitions towards some far distant good.

> It is a daily occurrence for men to act against what they know; they conceal it from themselves by turning their thoughts aside, so as to follow their passions. Otherwise we would not find people eating and drinking what they know will make them ill or even kill them. They would not neglect their affairs; they would not do what whole nations did, in certain ways, a few years ago.[46] The future and reasoning seldom strike as forcefully as do the present and the senses. ... Unless we resolve firmly to keep our minds on true good and true evil, so as to pursue the one and avoid the other, we find ourselves carried away, and the most important needs of this life are treated in the same way as heaven and hell are, even by their truest believers:
>
> We sing of this and praise it,
> We tell and hear about it,
> We write of it and read of it ...
> And then we do without it.
>   (*New Essays*: A VI vi 94–95;
>   RB 94–95)[47]

Given our confused perceptions of future events, we find it relatively easy to discount or downplay the threat of distant future harm in our present deliberations. Local communities are far more likely to take action against proposed developments whose effects are expected to be seen in the near future than they are to take action against proposals whose effects will harm future generations but not their own. For years, warnings of future climate change were either ignored or, when this was not an option, challenged and rebutted, the growing body of scientific evidence questioned. Little or no action was taken either individually or collectively. Only since anthropogenic climate change has become a present reality have we woken up to the crisis and begun to take action.

But is it inevitable that we underestimate distant future harms and overvalue near future goods? Will humans, for instance, always prefer gas-guzzling, status-conferring, convenient forms of transportation in the present even when they suspect that in doing so they are endangering the future existence of humanity and other living beings? An affirmative answer would be unduly pessimistic and not at all in keeping with Leibnizian optimism. The possibility of progressive advancement of ethical awareness is open. By Leibniz's account, we cannot make our perceptions of future events more distinct relative to our perceptions of the present: our perceptions of future events will in time become more distinct, but then they will no longer be events that are yet to happen in the future, but events that are happening in what will then be the present. Nevertheless, there are two reasons for optimism worth considering here. The first is that the confused nature of the perceptions that we have of the future has a positive side to it; the second is that successful attempts to make our perceptions in the present more distinct can be expected to have beneficial effects on the future.

Confused perceptions play a key role in Leibniz's metaphysics. Were our perceptions not confused, we would not perceive our own organic bodies through which we perceive the external world and without which we would 'be divorced from the universal bond, like deserters from the general order' (*Considerations on Vital Principles and Plastic Natures*: GP VI 546; L 590), an order that extends from the past, through the present, and into the future. Confused perceptions play an equally important role in facilitating our imaginative attempts to empathically 'put ourselves in the place of the other'.[48] In chapter 7, we examined compassionate empathic communication in terms of empathy with others coexisting in the present, but there is no reason why such communication should be restricted to those in our immediate purview. Using our imaginations, we can strive to develop empathic relations to those not yet born who have not yet started the developmental progression.[49] We can try as far as we can to put ourselves in the place of future others, imagining for instance what it would be like for them to live in a world in which sea levels have risen dramatically, land mass is reduced, erratic and extreme weather conditions lead to crop failures, and polluted seas no longer sustain life. And through such empathic communication with future beings, we might extend the compassion we feel for those with whom we coexist also to those who will succeed us.

Despite the advantages of having confused perceptions, striving to make our perceptions of the present as distinct as possible has its own advantages in relation to the future. As we noted earlier, the 'lineaments of the future' are already present in each preceding moment.[50] How we perceive and act now in part influences how the future will be. The perfections and imperfections of our souls are mirrored in the perfections and imperfections of our bodies and in the bodies and souls of others, past, present and future. The image in us of the perfections of others 'causes some of this perfection to be implanted and aroused within ourselves' (*On Wisdom*: GP VII 86; L 425).[51] Conversely, of course, the image of our own perfection must also implant and arouse some of that perfection in all others who perceive it. Given the perceptual mirroring of all by all, it follows that all creatures benefit in some way from the perfections of any one. Perfecting our own souls by increasing the distinctness of our perceptions and developing rational appetites towards the good is the most sensible strategy for ensuring that our minds and our bodies are operating effectively in accord with God's vision of the best possible world and its progression towards greater perfection. The more distinct are our perceptions, the more likely we are to perceive what is truly good as opposed to being attracted to things that merely appear to be good but are not really so. Finding a cut-price T-shirt in the bargain barrel at the local supermarket may seem like a good thing, but if its actual cost includes the perpetuation of the exploitation of workers in developing countries, it is not a true good. True goods do not cause unnecessary suffering to others. The wise are aware of the need to evaluate goodness in terms of the wellbeing or flourishing of all creatures that comprise the world. Living well is exemplified in humans in the virtue of the wise. It is the wise who may be said to be truly happy, enjoying a true 'serenity of spirit'[52] and deriving pleasure from the love of God and the contemplation of the beauty and perfection of creation. With distinct perceptions of what is truly good, the wise realize that benevolence towards all (where 'all', in our amendment of Leibniz, really does mean 'all', human and nonhuman alike) is also in their own best interest for they themselves take pleasure in the perception and promotion of the perfection, wellbeing, happiness and specific goods of others.[53] Wise love of God and universal benevolence towards all God's creation is not restricted to love of beings now coexisting in the same space; it extends also to those who participate in the same temporal flow (i.e. love towards all beings in each of their past, present and future states). Universal benevolence or the just charity of the wise embraces beings in the future as well as in the present. The onto- and bioegalitarianisms proposed in chapter 4 apply as much to future generations as they do to present ones. Those who have yet to develop have as much right to receive our just and fair treatment as do those with whom we coexist in the present.

Of course, even the wisest of humans do not always judge aright. As we have seen, no finite being has distinct perceptions that are free from confusion altogether. Unanticipated consequences can wreck even the best-laid plans. Acts that stem from the best of loving intentions can have seriously damaging outcomes, just as those that come from hatred and fear can produce unexpected goods. But whatever the actual outcome, by Leibniz's account, the traces of the original intention will be retained as traces in future events and experiences. The present vision and intentions will not be erased, but will continue into the future, being represented across the temporal divide in souls' and entelechies' future perceptions. The importance, therefore, of living well in the present is clear. It remains the case that our most effective – and indeed our only – means of influencing the future for better rather than for worse is to strive as best we can to make our present appetitions, perceptions and actions as good, morally and aesthetically, as they can possibly be. In this lies our own virtue and happiness. To become wise and to act in accordance with that wisdom is the greatest contribution we can make to the whole. Distinctly perceiving the perfection (as order and variety) of the universe, while also appreciating that it is on a path towards ever greater moral perfection, generates love towards all creation, happiness in the beholder, trust in the Creator, and an optimistic, hope-filled vision to hold onto when events appear to be travelling in the opposite direction.

In this regard, Voltaire's proposal that we focus on tending the garden has both literal and metaphorical significance. Towards the end of *Candide*, Pangloss and the other protagonists are offered hospitality by an old 'honest Turk' who lives with his family on a 20-acre farm. He tells his guests that he contents himself with the cultivation of his land and that their labours protect him and his family against 'three great evils, idleness, vice, and want' (Voltaire 2009 [1759], 132). Arguably, the honest Turk is the happiest character in *Candide*. Certainly, this is the opinion of Candide himself. As the friends contemplate the

Turk's situation on their journey home to their own small farm, Candide admits to finding the Turk's contented family life far preferable to the trappings of wealth and privilege and concludes that they too 'must take care of [their] garden' (Voltaire 2009 [1759], 133). In this light, Voltaire's repetition of the advice at the very end of the tale is not the mocking parody of Panglossian–Leibnizian optimism that it is often taken to be. On the contrary, Voltaire is making a serious point in a light-hearted manner when he writes:

> Pangloss [Leibniz] used sometimes to say to Candide, 'there is a concatenation of events in this best of all possible worlds, for if you had not been kicked out of a magnificent castle on account of miss Cunegonde, if you had not been thrown into the inquisition, if you had not rambled all over America on foot, if you had not run the Baron through the body, if you had not lost all your fine sheep of El Dorado, you would not be here to eat preserved citrons and pistachio nuts'.
> 'All that is very well', answered Candide, 'but let us take care of our garden'.
>
> (Voltaire 2009 [1759], 133)

Active participation in the processes of life and the creation of natural beauty is at once relaxing, energizing and stress-relieving, but at a deeper level, the good gardener pays close attention to the fine details that alert him or her to the distinctive needs of individual plants, understanding that 'one size does not fit all' and knowing that a key factor to success is the intimate relation between plant and environment. The good gardener also understands that 'there is a time for everything, and a season for every activity under the heavens' (Ecclesiastes 3:1), knowing instinctively when to prune and when to plant; when to work and when to wait; when to harvest and when to enjoy. The vital preparation of the land and the eventual planting of seeds or young plants incorporate both hope and expectation of their flowering and fruiting. The forester may never see the sapling trees emerge in the fullest of their glory, but will take pleasure from the knowledge that others in the future will enjoy their shade and shelter and health-giving benefits. In these and other ways, as Cooper has argued, gardening encourages virtues such as patience and humility (Cooper 2006).

Voltaire was surely right to advise the tending of the garden. To work in the garden, to savour its fruits, to take pleasure from the beauty of creation, to enjoy the company of friends and family, these are things too often undervalued but they are precisely the ingredients needed for the world to be perfect or as perfect as it can be. In a metaphorical sense, the whole world is a garden – a place of life and growth. Our own environments, whether literally or metaphorically, are gardens. In the words of the familiar hymn, 'All the world is God's own field' (Henry Alford).[54] Taken together, our own garden-environments are parts of the garden that is the whole of creation. Whether these are actual gardens or the metaphorical gardens that are our city homes, workplaces and local communities, our own garden-environments are not just physical places, but relational psychical places of value within a global relational psychical value space (chapter 6).

Voltaire's honest Turk does not concern himself with grander affairs of state,[55] but though he is right to concentrate on the local matters of his own farm, in this interconnected world in which the effects of the smallest activity ripples through the whole and will eventually rebound,[56] Schumacher's injunction to act locally while thinking globally still rings true. As we tend our own gardens – as we care for those in our immediate environment, striving for perfection and beauty – we need to keep an eye also on all that is spatially as well as temporally distant. The interconnectedness and mirroring of all parts of the universe entail that the harming of any one creature brings harm to all. Our relational value is a function of how well we direct our efforts to the care of all living things, how well we strive to promote the wellbeing of all creatures and inanimate things and work to bring to fruition the best possible world (chapter 5). This includes taking action to ensure that not only our literal gardens, but also our homes, work places, local communities are places of beauty and harmony that contribute as parts to the general beauty, harmony and perfection of the whole. In these endeavours, Leibniz's relational, dynamic, panpsychist pluralism may well fulfil Leibniz's wish that his philosophy would prove useful, though he himself could only have insensibly perceived that it might serve today as a foundation of an ecological philosophy that motivates hope-filled activity aimed at recognizing and increasing the perfection of the best of all possible worlds.

Notes

1 Similar statements abound. For instance, 'I have said more than once, that I hold space to be something merely relative, as time is; that I hold it to be an order of coexistences, as time is an order of successions' (Leibniz's Third Letter to Clarke: GP VII 363; LC 25–26); '*Time is the order of existence of those things which are not simultaneous*' (*The Metaphysical Foundations of Mathematics*: GM VII 17; L 666).

2 The instantaneous relational experiences had by monadic souls or the instantaneous relational states of bodies are the relational qualities upon which temporal relations among existing beings are founded, just as spatial relations are founded upon, for instance, the relational qualities of lines L and M (chapter 6). As such, instants are not themselves parts of time, but its foundational requisites (see Leibniz's Fifth Letter to Clarke: GP VII 402; LC 72–73).

3 Space and time are 'things true but ideal, like numbers' (to Conti, November or December 1715: LC 185).

4 Arthur (2014, esp. 159–165) offers an intelligent and informed reading of Leibniz on the actuality of relational time as the 'order of successive things'.

5 However, as Arthur observes, this would require that the motion of bodies be taken into consideration and the development of a theory of space-time (Arthur 2014, 164–165).

6 If God had created only disembodied monads, there would be time, but no space. Physical relational space depends upon bodies existing in relation to each other. Even in the absence of bodies, however, relational time is possible.

7 At this point, one might ask why the internal relations between God's perceptions do not themselves constitute time that is prior to the creation of finite substances. Leibniz's God, however, is independent of time. God's perceptions do not unfold one

by one in sequential temporal order. The eternity of God is not the durational time of finite beings (Leibniz's Fifth Letter to Clarke: GP VII 415–416; LC 90).

8 See also chapter 5, p. 100.

9 *Treatise* 1.1.1.

10 The majority of our present perceptual states are composed of confused and insensible perceptions (chapter 7, pp. 135–136). Being perceived confusedly or insensibly is therefore not in itself the reason why some events are past or future. Rather the confusion or obscurity of the perception must be considered in conjunction with its place in the sequence of perceptions as a whole.

11 See also Nita 2008, 16.

12 See chapter 7, pp. 135–137.

13 Leibniz's insights were, however, taken up and improved upon in the centuries that followed. See Arthur 2014, 165.

14 Chapter 7, pp. 142–143.

15 The body can only be an 'entire effect' if it is capable of holding in one unitary point all of the causes that had led to its being the way it currently is. Thus, only organic bodies endowed with dominating souls or entelechies are properly 'entire effects' for it is only by the soul's or entelechy's unifying act of representation that the physical impressions that have been made on each of the infinitely many parts of its body are united in the present single perceptual state (Phemister 2015, 137–140). Consequently, only matter that is imbued throughout with souls or entelechies can play a role in the mechanistic philosophy. This conclusion thereby reinforces Leibniz's pluralist panpsychism.

16 Leibniz's intention that his brief history of the earth, the *Protogaea*, should serve as the introduction to his history of the House of Brunswick (Leibniz 2008b, xiii) is a rather mischievous product of his commitment to the principle of the interconnectedness of all things and the uninterrupted causal chains that link the past to the present.

17 Chapter 7, pp. 138–142.

18 Death, in Leibniz's opinion, is not the end of existence, but rather is a dramatic transformation of the corporeal substance in which its organic body contracts into an invisible physical point and its soul sinks into a stupor-like state. See chapter 4, note 11 (p. 89). From this, it follows that not only do the effects of present actions continue into the future, but each living thing continues to exert an influence even after death.

19 Or, as he says in *Whether the World Increases in Perfection*, 'no perfect oblivion is granted to souls' (Grua I 95; ST 197). No soul or entelechy is separated from its body even on death, but Leibniz believed that God treated rational souls differently from those whose bodies on death are assimilated in minute form into the general mass of extended matter (*New System*: GP IV 480–481; AG 140–141). When rational creatures die, he proposes, their souls might 'retain a subtle body organized after its own manner' of a kind that scholastic theologians had attributed to angels (*Reflections on the Doctrine of a Single Universal Spirit*: GP VI 533; L 556–557). Leibniz's comments raise some intriguing questions about the 'places' our ancestors might inhabit. For instance, what relations do their subtle bodies have to our gross bodies? If these relations are formative of the places and space we inhabit, does this mean we share the same physical and psychical places with our ancestors? In short, do Leibniz's views entail that the ghosts of our ancestors are presently all around us?

20 E.g. see chapter 4, pp. 75–78 and chapter 6, pp. 118–121.

21 E.g. *Monadology* §§4–6: GP VI 607; AG 213.

22 *Discourse on Metaphysics* §34: GP IV 459; AG 65. See also *Monadology* §§29, 30: GP VI 611–612; AG 217 and *Principles of Nature and of Grace* §5: GP VI 600–601; AG 209.

23 *Principles of Nature and of Grace* §5: GP VI 600; AG 208. See also, *Monadology* §28: GP VI 611; AG 216.

24 ... by one whose perceptions are sufficiently distinct ....

25 An individual's degree of perfection may be said to be increasing over a period of time provided any setbacks do not take the individual below the previous lowest point: 'If any substance progresses in perfection to infinity, whether directly or by interposed regressions, it is necessary that it can now be assigned a maximum degree of perfection below which it will never descend in the future, and in turn, after some time has elapsed, another degree of perfection can be assigned which is greater than the former' (*Whether the World Increases in Perfection*: Grua 1 95; ST 196). Stating this in terms of the clarity and distinctness of rational souls' perceptions, we can say that a soul's perceptions are increasing in perfection overall so long as they never lose the clarity that they had reached before the point at which their perceptions started to become more distinct again.

26 Gale 1976, 74–75. See Strickland 2006, 115.

27 It is not obvious why this should be so, for it is surely possible that there be some final causes that lead to increased perfection in the parts, without there being any increase in the perfection of the whole. The implication would, however, follow if by 'final cause' Leibniz is referring to the final cause that is God's vision of the unfolding of the whole.

28 Chapter 5, p. 100.

29 Chapter 6, pp. 117–121 and chapter 7, pp. 138–140.

30 See chapter 6, p. 121.

31 See pp. 160–161.

32 For a dissenting voice, see Strickland 2006, 120, 138–139n5.

33 The distinctness of a monad's perceptions results from the activity of its appetitive force, which is itself a modification of the monad's primitive active force. Its primitive active force, however, is also modified as the derivative active force of its organic body (chapter 2, p. 40). The amount of derivative active force in the world remains constant from moment to moment. Consequently, one would expect that the amount of appetitive force across the world must also remain constant, such that increases in the appetitive forces of some monads must be balanced by corresponding decreases in the appetitive forces of others. As Leibniz himself admits in *Whether the World Increases in Perfection* 'some substances cannot continually increase in perfection without others continually decreasing in perfection' (Grua I 95: ST 196, cited above, p. 163). In the passage from *On the Ultimate Origination of Things* to which this note refers, Leibniz hints that the cultivation and moral perfectioning of some substances is counterbalanced by the infinity of monads in 'the abyss'. Their perfectibility has not yet begun and given that there are infinitely many of them, no matter how many start to progress, there will always be others whose decline will keep the overall balance of forces intact. See also Phemister 2006.

34 Chapter 4, p. 70.

35 Chapter 2, pp. 35–36.

36 *Discourse on Metaphysics* §8: GP IV 433; AG 41. Of course, our perceptions can never be so distinct that we can have complete knowledge and understanding of ourselves and the co-constituting relations to others through which our individual identities are formed and which locate us in this world rather than in any other (see chapter 5, pp. 98–99).

37 *Ethics* I, P29 and P33.

38 Chapter 5, pp. 98–99.

39 Nor was it absolutely necessary that God created this world rather than any of the others. He could have chosen to create a different, less perfect world, although he would never actually have done so for God always chooses what he knows to be the best, just as we always choose what we believe to be the best. In both cases, teleological reasons that involve calculations as to the best courses of action underpin the choices made, but as Leibniz was fond of saying, these reasons merely 'incline without necessitating'. The agent remains free to choose otherwise then they do

(*Remarks upon M. Arnauld's Letter* ...: GP II 46; LA 50, *New Essays*: A VI vi 178; RB 178). For further discussion of these issues, see Phemister 2005, ch. 10 and 2007.

40 Phemister 1991.
41 Chapter 4, p. 77.
42 Alastair McIntosh finds less reason to be optimistic. He writes: 'There is a slow urgency about what humankind's levels of consumption are doing to the earth. Slow, in that it is difficult to register it in ways that can trigger radical action on political timescales. But urgent, in that the ratchet is tightening especially if we have any care at all for future generations. The tipping points of no return show signs of slipping. That is why I have been forced to abandon optimism and seek recourse in hope' (McIntosh 2008, 249).
43 Voltaire 2009 [1759].
44 Although the process has its detractors, most agree that it was at least partially successful in restoring peace in South Africa.
45 See above, p. 158.
46 Possibly a reference to the Thirty Years War, which ended in 1648.
47 Using similar reasoning based on the force and vivacity of present impressions on our senses, Hume came to the same conclusion. See *Treatise* 3.2.7.
48 Chapter 7, p. 148.
49 *On the Ultimate Origination of Things*: GP VII 308; AG 155.
50 See above, p. 159, and chapter 7, p. 143.
51 See chapter 7, p. 139.
52 Chapter 6, p. 125.
53 Chapter 6, pp. 119–121.
54 Hymn no. 123 in *Singing the Faith* (Ferndown, Dorset: Hymns Ancient & Modern, 2011).
55 Voltaire 2009 [1759], 132.
56 Chapter 7, p. 146.

Bibliography

Abram, David (1996) *The Spell of the Sensuous: Perception and Language in a More-Than-Human World*, New York: Random House.

Adams, Robert Merrihew (1994) *Leibniz: Determinist, Theist, Idealist*, Oxford: Oxford University Press.

Ansell Pearson, Keith (2000) 'Nietzsche's Brave New World of Force: On Nietzsche's 1873 "Time Atom Theory" Fragment and the Influence of Boscovich on Nietzsche', *Journal of Nietzsche Studies* 20, 5–33.

Antognazza, Maria Rosa (2009) *Leibniz: An Intellectual Biography*, Cambridge: Cambridge University Press.

Aristotle (2004) *The Nicomachean Ethics*, trans. J. A. K. Thomson, ed. Hugh Tredennick, London: Penguin.

Armstrong-Buck, Susan (1986) 'Whitehead's Metaphysical System as a Foundation for Environmental Ethics', *Environmental Ethics* 8:3, 241–259.

Arthur, Richard T. W. (2014) *Leibniz*, Cambridge: Polity.

Barber, William Henry (1985 [1955]) *Leibniz in France: From Arnauld to Voltaire*, New York and London: Garland Publishing. Repr., orig. publ., Clarendon Press, Oxford, 1955.

Basile, Pierfrancesco (2006) '"All Monads have Windows": James Ward and the Reception of Leibniz's Theory of Monads in British Idealism', in Herbert Breger, Jürgen Herbst and Sven Erdner (eds), *Einheit in der Veilheit*, Proceedings of the VIII Internationaler Leibniz-Kongress, Hannover, 24–29 July 2006, 3 vols, Hannover: Gottfried-Wilhelm-Leibniz-Gesellschaft, vol. 1, 29–36.

Basile, Pierfrancesco (2009) *Leibniz, Whitehead and the Metaphysics of Causation*, Basingstoke: Palgrave Macmillan.

Baumgarten, Alexander (2013) *Alexander Baumgarten, Metaphysics: A Critical Translation with Kant's Elucidations, Selected Notes, and Related Materials*, ed. and trans. Courtney D. Fugate and John Hymers, London: Bloomsbury Academic.

Beiser, Frederick C. (2009) *Diotima's Children: German Aesthetic Rationalism from Leibniz to Lessing*, Oxford: Oxford University Press.

Beiser, Frederick C. (2013) *Late German Idealism: Trendelenburg and Lotze*, Oxford: Oxford University Press.

Beiser, Frederick C. (2015) 'Herbart's Monadology', *British Journal for the History of Philosophy* 23:6, 1056–1073.

Bekoff, Marc (2007) *The Emotional Lives of Animals: A Leading Scientist Explores Animal Joy, Sorrow, and Empathy – And Why They Matter*, Novato, CA: New World Library.

Bekoff, Marc (2012) 'Learning from Animals', *Resurgence* 271, 28–31.

Bennett, Jane (2004) 'The Force of Things: Steps towards an Ecology of Matter', *Political Theory* 32:3, 347–372.

Bennett, Jane (2010) *Vibrant Matter: A Political Ecology of Things*, Durham and London: Duke University Press.

Bennett, Jane (2012) 'Artistry and Agency in a World of Vibrant Matter', paper presented at the New School in New York, <http://www.youtube.com/watch?v=q607Ni23QjA>, accessed 8 December 2012.

Bennett, Jonathan (1984) *A Study of Spinoza's 'Ethics'*, Indianapolis: Hackett Publishing Co.

Bolton, Martha (2011) 'Leibniz's Theory of Cognition', in Brandon Look (ed.), *The Continuum Companion to Leibniz*, London: Continuum, 136–158.

Bradley, Francis Herbert (1920 [1893]) *Appearance and Reality: A Metaphysical Essay*, 2nd edn, 7th impr., London: George Allen & Unwin.

Brady, Emily and Pauline Phemister (eds) (2012) *Human–Environment Relations: Transformative Values in Theory and Practice*, Dordrecht: Springer.

Bristow, Tom (2012) 'Toward History and the Creaturely: Language and the Intertextual Literary Value Space in Jonathan Safran Foer's *Eating Animals*', in Emily Brady and Pauline Phemister (eds), *Human–Environment Relations: Transformative Values in Theory and Practice*, Dordrecht: Springer, 69–83.

Broad, Charles Dunbar (1979 [1975]) *Leibniz: An Introduction*, ed. Casimir Lewy, corr. repr., Cambridge: Cambridge University Press.

Brown, Sam P. and Rufus A. Johnstone (2001) 'Cooperation in the Dark: Signalling and Collective Action in Quorum-Sensing Bacteria', *Proceedings of the Royal Society Biological Sciences* 268:1470, 961–965.

Butler, Edward (2005) Review of *Spinoza and Deep Ecology: Challenging Traditional Approaches to Environmentalism*, by Eccy de Jonge (Ashgate 2004), *Metapsychology* 9:45, <http://metapsychology.mentalhelp.net/poc/view_doc.php?type=book&id=2893&cn=394>, accessed 11 March 2014.

Carlson, Allen (1984) 'Nature and Positive Aesthetics', *Environmental Ethics* 6:1, 5–34.

Carlson, Allen (2000) *Aesthetics and the Environment: The Appreciation of Nature, Art and Architecture*, London: Routledge.

Carlson, Allen (2004) 'Appreciation and the Natural Environment', in Allen Carlson and Arnold Berleant (eds), *The Aesthetics of Natural Environments*, Peterborough, ON: Broadview Press, 63–75.

Carr, Herbert Wildon (1922) *A Theory of Monads: Outlines of the Philosophy of the Principle of Relativity*, London: Macmillan.

Carr, Herbert Wildon (1930) *Cogitans Cogitata*, London: Favil Press.

Cassirer, Ernst (2009 [1951]) *The Philosophy of the Enlightenment*, trans. Fritz C. A. Koelln and James P. Pettegrove, Princeton: Princeton University Press.

Clarke, Desmond M. (2003) *Descartes's Theory of Mind*, Oxford: Clarendon Press.

Conger, George Perrigo (1922) *Theories of Macrocosms and Microcosms in the History of Philosophy*, New York: Columbia University Press.

Conway, Anne (1996) *The Principles of the Most Ancient and Modern Philosophy*, ed. Allison P. Coudert and Taylor Corse, Cambridge: Cambridge University Press.

Cooper, David E. (2006) *A Philosophy of Gardens*, Oxford: Clarendon Press.

Copleston, Frederick (1963) *A History of Philosophy*, vol. 7: *Fichte to Nietzsche*, London: Burns & Oates.

Cornforth, Daniel M., Roman Popat, Luke McNally, James Gurney, Thomas C. Scott-Phillips, Alasdair Ivens, Stephen P. Diggle and Sam P. Brown (2014) 'Combinatorial Quorum Sensing Allows Bacteria to Resolve Their Social and Physical Environment', *Proceedings of the National Academy of Sciences of the United States of America* 111:11, 4280–4284.

Cottingham, John (1998) 'Descartes' Treatment of Animals', in John Cottingham (ed.), *Descartes*, Oxford: Oxford University Press, 225–233.

Courchamp, Frank, Elena Angulo, Philippe Rivalan, Richard J. Hall, Laetitia Signoret, Leigh Bull and Yves Meinard (2006) 'Rarity Value and Species Extinction: The Anthropogenic Allee Effect', *PLoS Biology* 4:12, 2405–2410. <http://www.plosbiology.org/article/fetchObject.action?uri=info%3Adoi%2F10.1371%2Fjournal.pbio.0040415&representation=PDF>.

Couturat, Louis (1901) *La logique de Leibniz, d'après des document inédits*, Paris: Félix Alcan.

Cover, Jan A. and John O'Leary-Hawthorne (1999) *Substance and Individuation in Leibniz*, Cambridge: Cambridge University Press.

Crist, Eileen (2004) 'Can an Insect Speak? The Case of the Honeybee Dance Language', *Social Studies of Science* 34:1, 7–43.

Cristin, Renato (1998) *Heidegger and Leibniz: Reason and the Path*, Dordrecht: Springer.

Crutzen, Paul J. and Eugene F. Stoermer (2000) 'The "Anthropocene"', *Global Change Newsletter* 41, 17–18.

Davis, Martin (2000) *The Universal Computer: The Toad from Leibniz to Turing*, London: Norton.

de Jonge, Ecce (2004) *Spinoza and Deep Ecology: Challenging Traditional Approaches to Environmentalism*, Aldershot: Ashgate.

Delafield-Butt, Jonathan T. (2014) 'Process and Action: Whitehead's Ontological Units and Perceptuomotor Control Units', in Spyridon Koutroufinis (ed.), *Life and Process*, Berlin and Boston: De Gruyter Ontos, 133–156.

Delafield-Butt, Jonathan T., Gert-JanPepping, Colin D.McCaig and David N. Lee (2012) 'Prospective Guidance in a Free-Swimming Cell', *Biological Cybernetics* 106, 283–293.

Deleuze, Gilles (1980) 'Leibniz: Cours Vincennes – 15/04/1980', in Gilles Deleuze, *Les cours de Gilles Deleuze*, <http://www.webdeleuze.com/php/texte.php?cle=50&groupe=Leibniz&langue=2>, accessed 12 July 2015.

Deleuze, Gilles (1988) *Spinoza: Practical Philosophy*, trans. Robert Hurley, San Francisco: City Lights Books.

Della Rocca, Michael (1996) *Representation and the Mind–Body Problem in Spinoza*, Oxford: Oxford University Press.

Descartes, René (1964–76) *Oeuvres de Descartes*, ed. Charles Adam and Paul Tannery, 12 vols, Paris: Vrin.

Descartes, René (1985–91) *The Philosophical Writings of Descartes*, 3 vols, ed. and trans. John Cottingham, Robert Stoothoff and Dugald Murdoch, vol. 3 with Anthony Kenny, Cambridge: Cambridge University Press.

DeSouza, Nigel (2012) 'Leibniz in the Eighteenth Century: Herder's Critical Reflections on the *Principles of Nature and Grace*', *British Journal for the History of Philosophy* 20:4, 773–795.

di Bella, Stefano (2005) *The Science of the Individual: Leibniz's Ontology of Individual Substance*, Dordrecht: Springer.

di Giovanni, George (2011) 'The New Spinozism', in Alison Stone (ed.), *The Edinburgh Critical History of Nineteenth-Century Philosophy*, Edinburgh: Edinburgh University Press, 13–28.

Dunham, Jeremy (Forthcoming) 'A Universal and Absolute Spiritualism: Maine de Biran's Leibniz', in Darian Meacham and Joseph Spadola (eds), *The Relationship between the Physical and Moral in Man: The Philosophy of Maine de Biran*, London: Bloomsbury Academic.

European Communities (2005) *EUR 21796–Synthetic Biology–Applying Engineering to Biology: Report of a NEST High-Level Expert Group*, report compiled by Katia Van Compernolle and Philip Ball under the chairmanship of Luis Serrano, Luxembourg: Office for Official Publications of the European Communities.

Fichant, Michel (2004) 'L'Invention métaphysique', in Michel Fichant (ed.), *'Discours de métaphysique' suivi de 'Monadologie' et autres textes*, Paris: Gallimard, 7–140.

Foer, Jonathan Safran (2010) *Eating Animals*, London: Hamish Hamilton.

Foster, Ryan J. (2011) *The Creativity of Nature: The Genesis of Schelling's Naturphilosophie, 1775–1799*, Ann Arbor, MI: Proquest, UMI Dissertation Publishing.

Frost, Samantha (2005) 'Hobbes and the Matter of Self-Consciousness', *Political Theory* 33:4, 495–517.

Gale, George (1976) 'On What God Chose: Perfection and God's Freedom', *Studia Leibnitiana* 8:1, 69–87.

Garber, Daniel (2004) 'Leibniz and Fardella: Body, Substance, and Idealism', in Paul Lodge (ed.), *Leibniz and His Correspondents*, Cambridge: Cambridge University Press, 123–140.

Garber, Daniel (2009) *Leibniz: Body, Substance, Monad*, Oxford: Oxford University Press.

Gaukroger, Stephen (1995) *Descartes: An Intellectual Biography*, Oxford: Oxford University Press.

Goto Collins, Reiko (2012) *Ecology and Environmental Art in Public Place: Talking Tree; Won't You Take a Minute and Listen to the Plight of Nature?*, PhD thesis, Robert Gordon University, Aberdeen.

Gunter, Pete A. Y. (2006) 'Whitehead and Environmental Education', in George Allan and Malcolm D. Evans (eds), *A Different Three Rs for Education: Reason, Relationality, Rhythm*, Amsterdam: Rodopi, 75–86.

Gutting, Gary (2001) *French Philosophy in the Twentieth Century*, Cambridge: Cambridge University Press.

Guyer, Paul (2014) *A History of Modern Aesthetics*, 3 vols, Cambridge: Cambridge University Press.

Hailwood, Simon (2000) 'The Value of Nature's Otherness', *Environmental Values* 9:3, 353–372.

Haserot, Francis S. (1972) 'Spinoza's Definition of Attribute', in S. Paul Kashap (ed.), *Studies in Spinoza: Critical and Interpretive Essays*, Berkeley, CA: University of California Press, 28–42.

Hatfield, Gary (1998) 'The Cognitive Faculties', in Daniel Garber and Michael Ayers (eds), *The Cambridge History of Seventeenth-Century Philosophy*, 2 vols, Cambridge: Cambridge University Press, vol. 2, 953–1002.

Herder, Johann Gottfried (2002) *Philosophical Writings*, ed. and trans. Michael N. Forster, Cambridge: Cambridge University Press.

Hostler, John (1975) *Leibniz's Moral Philosophy*, New York: Barnes & Noble.

Hume, David (1975) *Enquiries concerning Human Understanding and concerning the Principles of Morals*, ed. Lewis Amherst Selby-Bigge, rev. Peter Harold Nidditch, 3rd edn, Oxford: Clarendon Press.

Hume, David (2000) *A Treatise of Human Nature*, ed. David Fate Norton and Mary J. Norton, Oxford: Oxford University Press.

Iltis, Carolyn (1971) 'Leibniz and the *Vis Viva* Controversy', *Isis* 62:1, 21–35.

Ishiguro, Hidé (1990 [1972]) *Leibniz's Philosophy of Logic and Language*, 2nd edn, Cambridge: Cambridge University Press.

James, Simon (2011) 'For the Sake of a Stone? Inanimate Things and the Demands of Morality', *Inquiry: An Interdisciplinary Journal of Philosophy* 54:4, 384–397.

Jordan, George Jefferis (1927) *The Reunion of the Churches: A Study of G. W. Leibnitz and His Great Attempt*, London: Constable & Co.

Kant, Immanuel (1900–) *Immanuel Kant's Gesammelte Schriften*, 29 vols, ed. Preußischen Akademie de Wissenschaften, Berlin: Akademie Verlag.

Kant, Immanuel (2001) *Critique of the Power of Judgment*, ed. Paul Guyer, trans. Paul Guyer and Eric Matthews, Cambridge: Cambridge University Press.

Kant, Immanuel (2012) *Natural Science*, ed. and trans. Eric Watkins, Cambridge: Cambridge University Press.

Kuehn, Manfred (2001) 'Kant's Teachers in the Exact Sciences', in Eric Watkins (ed.), *Kant and the Sciences*, Oxford: Oxford University Press, 11–27.

Kuhn, Thomas (1970 [1962]) *The Structure of Scientific Revolutions*, 2nd edn, Chicago: University of Chicago Press.

Kulstad, Mark (1977) 'Leibniz's Conception of Expression', *Studia Leibnitiana* 9:1, 55–76.

Kulstad, Mark (1980) 'A Closer Look at Leibniz's Alleged Reduction of Relations', *Southern Journal of Philosophy* 18:4, 417–432.

Kulstad, Mark (1982) 'Some Difficulties in Leibniz's Definition of Perception', in Michael Hooker (ed.), *Leibniz: Critical and Interpretive Essays*, Manchester: Manchester University Press, 65–78.

Kulstad, Mark (2006) 'Leibniz on Expression: Reflections after Three Decades', in Herbert Breger, Jürgen Herbst and Sven Erdner (eds), *Einheit in der Veilheit*, Proceedings of the VIII Internationaler Leibniz-Kongress, Hannover, 24–29 July 2006, 3 vols, Hannover: Gottfried-Wilhelm-Leibniz-Gesellschaft, vol. 1, 413–419.

Lærke, Mogens (2011) 'A Conjecture about a Textual Mystery: Leibniz, Tschirnhaus and Spinoza's "Korte Verhandeling"', *Leibniz Review* 21, 33–68.

Lærke, Mogens (2015) 'Five Figures of Folding: Deleuze on Leibniz's Monadological Metaphysics', *British Journal for the History of Philosophy* 23:6, 1192–1213.

Leibniz, Gottfried Wilhelm (1765) *Oeuvres philosophiques latines et françoises de feu Mr de Leibnitz, tirées de ses manuscrits qui se conservent dans la Bibliothèque Royale à Hanovre*, ed. Rudolf Erich Raspe, Amsterdam and Leipzig: chez Jean Schreuder.

Leibniz, Gottfried Wilhelm (1768) *Opera omnia: Nunc primum collecta, in classes distributa, praefationibus et indicibus exornata*, ed. Ludovici Dutens, 6 vols (in 7), Geneva: apud fratres de Tournes.

Leibniz, Gottfried Wilhelm (1838–40) *Leibnitz's Deutsche Schriften*, ed. Gottschalk Eduard Guhrauer, 2 vols, Berlin: Veit.

Leibniz, Gottfried Wilhelm (1849–63) *Leibnizens mathematische Schriften*, ed. Carl Immanuel Gerhardt, 7 vols, Berlin: A. Asher; Halle: H. W. Schmidt.

Leibniz, Gottfried Wilhelm (1854) *Réfutation inédite de Spinoza par Leibniz*, ed. Alexandre Foucher de Careil, Paris: E. Brière.

Leibniz, Gottfried Wilhelm (1857) *Nouvelles lettres et opuscules inédits de Leibniz*, ed. Alexandre Foucher de Careil, Paris: Durand.

Leibniz, Gottfried Wilhelm (1864–84) *Die Werke von Leibniz gemäss seinem hanschriftlichen Nachlasse in der Königlichen Bibliothek zu Hannover*, ed. Onno Klopp, 11 vols, Hannover: Klindworth.

Leibniz, Gottfried Wilhelm (1875–90) *Die philosophischen Schriften von Gottfried Wilhelm Leibniz*, ed. Carl Immanuel Gerhardt, 7 vols, Berlin: Weidman; repr., Hildesheim: Olms, 1965.

Leibniz, Gottfried Wilhelm (1903) *Opuscules et fragments inédits de Leibniz: Extraits des manuscrits de la Bibliothèque royale de Hanovre*, ed. Louis Couturat, Paris: Alcan.

Leibniz, Gottfried Wilhelm (1923–) *Sämtliche Schriften und Briefe*, ser. 1—7, ed. Deutsche Akademie der Wissenschaften, Berlin: Akademie Verlag.

Leibniz, Gottfried Wilhelm (1948) *Textes inédits*, ed. Gaston Grua, 2 vols, Paris: Presses Universitaires de France.

Leibniz, Gottfried Wilhelm (1956) *The Leibniz–Clarke Correspondence*, trans. Henry Gavin Alexander, Manchester: Manchester University Press.

Leibniz, Gottfried Wilhelm (1967) *The Leibniz–Arnauld Correspondence*, ed. and trans. Haydn Trevor Mason, Manchester: Manchester University Press.

Leibniz, Gottfried Wilhelm (1969) *Philosophical Papers and Letters*, ed. and trans. Leroy E. Loemker, 2nd edn, Dordrecht: Reidel Publishing Co.

Leibniz, Gottfried Wilhelm (1973) *Philosophical Writings*, ed. George Henry Radcliffe Parkinson, trans. Mary Morris and George Henry Radcliffe Parkinson, rev. edn, London: J. M. Dent.

Leibniz, Gottfried Wilhelm (1981) *New Essays on Human Understanding*, ed. and trans. Peter Remnant and Jonathan Bennett, Cambridge: Cambridge University Press.

Leibniz, Gottfried Wilhelm (1985) *Theodicy: Essays on the Goodness of God, the Freedom of Man, and the Origin of Evil*, trans. E. M. Huggard, LaSalle, IL: Open Court.

Leibniz, Gottfried Wilhelm (1989) *Philosophical Essays*, ed. and trans. Roger Ariew and Daniel Garber, Indianapolis and Cambridge: Hackett Publishing Co.

Leibniz, Gottfried Wilhelm (1992) *De summa rerum: Metaphysical Papers, 1675–1676*, ed. and trans. George Henry Radcliffe Parkinson, New Haven: Yale University Press.

Leibniz, Gottfried Wilhelm (1997) *Leibniz's 'New System' and Associated Contemporary Texts*, ed. and trans. Roger S. Woolhouse and Richard Francks, Oxford: Clarendon Press.

Leibniz, Gottfried Wilhelm (2006) *The Shorter Leibniz Texts: A Collection of New Translations*, ed. and trans. Lloyd Strickland, London: Continuum.

Leibniz, Gottfried Wilhelm (2007) *The Leibniz–Des Bosses Correspondence*, ed. and trans. Brandon C. Look and Donald Rutherford, New Haven and London: Yale University Press.

Leibniz, Gottfried Wilhelm (2008a) *The Art of Controversies*, ed. and trans. Marcelo Dascal, with Quintín Racionero and Adelino Cardoso, Dordrecht: Springer.

Leibniz, Gottfried Wilhelm (2008b) *Protogaea: Gottfried Wilhelm Leibniz*, ed. and trans. Claudine Cohen and Andre Wakefield, Chicago and London: University of Chicago Press.

Leibniz, Gottfried Wilhelm (2011) *Leibniz and the Two Sophies: The Philosophical Correspondence*, ed. and trans. Lloyd Strickland, Toronto: Centre for Reformation and Renaissance Studies.

Leibniz, Gottfried Wilhelm (2013) *G. W. Leibniz: The Leibniz–De Volder Correspondence*, ed. and trans. Paul Lodge, New Haven and London: Yale University Press.

Leibniz, Gottfried Wilhelm (2014) *Leibniz's 'Monadology': A New Translation and Guide*, ed. and trans. Lloyd Strickland, Edinburgh: Edinburgh University Press.

L'Engel, Madeleine (1973) *A Wind in the Door*, New York: Square Fish.

Lloyd, Genevieve (1980) 'Spinoza's Environmental Ethics', *Inquiry* 23:3, 293–311.

Lloyd, Genevieve (1993 [1984]) *The Man of Reason; 'Male' and 'Female' in Western Philosophy*, 2nd edn, London: Routledge. Orig. publ.,London: Methuen.

Locke, John (1975) *An Essay concerning Human Understanding*, ed. Peter H. Nidditch, Oxford: Clarendon Press.

Locke, John (2008) *An Essay concerning Human Understanding*, ed. and abridged by Pauline Phemister, Oxford: Oxford University Press.

Lodge, Paul (2001) 'Leibniz's Notion of an Aggregate', *British Journal for the History of Philosophy* 9:3, 467–486.

Lodge, Paul (2015) 'Heidegger on the Being of Monads: Lessons in Leibniz and in the Practice of Reading the History of Philosophy', *British Journal for the History of Philosophy* 23:6, 1169–1191.

Lotze, Hermann (1888) *Microcosmus: An Essay concerning Man and His Relation to the World*, trans. Elizabeth Hamilton and E. E. Constance Jones, 3rd edn, 2 vols, Edinburgh: T & T Clark.

McCullough, Laurence B. (1996) *Leibniz on Individuals and Individuation: The Persistence of Premodern Ideas in Modern Philosophy*, Dordrecht: Kluwer.

McIntosh, Alastair (2008) *Hell and High Water: Climate Change, Hope, and the Human Condition*, Edinburgh: Birlinn.

Majumdar, J. K. (1929) 'The Idealism of Leibniz and Lotze', *Philosophical Review* 38:5, 456–468.

Malebranche, Nicolas (1979–92) *Oeuvres*, ed. and trans. Geneviève Rodis-Lewis, in collaboration with Germain Malbreil, 2 vols, Paris: Gallimard.

Malebranche, Nicolas (1997) *The Search after Truth: With Elucidations of The Search after Truth*, ed. and trans. Thomas M. Lennon and Paul J. Olscamp, Cambridge: Cambridge University Press.

Masson, Jeffrey Moussaieff and Susan McCarthy (1996) *When Elephants Weep: The Emotional Lives of Animals*, London: Vintage.

Mathews, Freya (1991) *The Ecological Self*, London: Routledge.

Mercer, Christia (2001) *Leibniz's Metaphysics: Its Origins and Development*, Cambridge: Cambridge University Press.

Merchant, Carolyn (1979) 'The Vitalism of Francis Mercury van Helmont: Its Influence on Leibniz', *Ambix* 26:3, 170–183.

Merchant, Carolyn (1989 [1980]) *The Death of Nature: Women, Ecology and the Scientific Revolution*, San Francisco: Harper & Row.

Merchant, Carolyn (1992) *Radical Ecology: The Search for a Liveable World*, London: Routledge.

Mesle, Robert C. (2008) *Process-Relational Philosophy: An Introduction to Alfred North Whitehead*, West Conshohocken, PA: Templeton Foundation Press.

Montaigne, Michel de (1987) *An Apology for Raymond Sebond*, ed. and trans. M. A. Screech, London: Penguin Classics.

Montaigne, Michel de (1993 [1991]) *The Complete Essays*, ed. and trans. M. A. Screech, London: Penguin Classics.

Morrison, James C. (1989) 'Why Spinoza Had No Aesthetics', *Journal of Aesthetics and Art Criticism* 47:4, 359–365.

Mugnai, Massimo (1992) *Leibniz's Theory of Relations*, Studia Leibnitiana Supplementa 28, Stuttgart: Franz Steiner.

Naaman Zauderer, Noa (2009) 'The Place of the Other in Leibniz's Rationalism', in Marcelo Dascal (ed.), *Leibniz: What Kind of Rationalist?*, Dordrecht: Springer, 315–327.

Nachtomy, Ohad (2007) *Possibility, Agency, and Individuality in Leibniz's Metaphysics*, Dordrecht: Springer.

Naess, Arne (1969) 'Freedom, Emotion, and Self-Subsistence: The Structure of a Small, Central Part of Spinoza's *Ethics*', *Inquiry: An Interdisciplinary Journal of Philosophy* 12:1–4, 66–104.

Naess, Arne (1973) 'The Shallow and the Deep, Long-Range Ecology Movement: A Summary', *Inquiry: An Interdisciplinary Journal of Philosophy* 16:1–4, 95–100.

Naess, Arne (1977) 'Spinoza and Ecology', *Philosophia* 7:1, 45–54.

Naess, Arne (1986) 'The Deep Ecological Movement: Some Philosophical Aspects', *Philosophical Inquiry* 8:1–2, 10–31. Rev. version, in Michael Zimmermann et al. (eds), *Environmental Philosophy: From Animal Rights to Radical Ecology*, Englewood Cliffs: Prentice-Hall, 1998; repr., Andrew Light and Holmes Rolston III (eds), *Environmental Ethics: An Anthology*, Oxford: Blackwell, 2003, 262–274.

Naess, Arne (1988) 'Deep Ecology and Ultimate Premises', *Ecologist* 18:4–5, 128–131.

Naess, Arne (1989) *Ecology, Community and Lifestyle: Outline of an Ecosophy*, trans. and revised by David Rothenberg, Cambridge: Cambridge University Press.

Naess, Arne and George Sessions (1984) 'Basic Principles of Deep Ecology', *Ecophilosophy* 6, 3–7, <http://environmentalethics.info/ecophilosophynewsletters/ecophilosophy6.pdf>, accessed 30 May 2015.

Nagel, Thomas (1974) 'What Is It Like to Be a Bat?', *Philosophical Review* 83:4, 435–450.

Nita, Adrian (2008) *La métaphysique du temps chez Leibniz et Kant*, Paris: L'Harmattan.

O'Briant, Walter H. (1980) 'Leibniz's Contribution to Environmental Philosophy', *Environmental Ethics* 2:3, 215–220.

O'Neill, John (1992) 'The Varieties of Intrinsic Value', *Monist* 75:2, 119–138.

Packard, Andrew and Jonathan T. Delafield-Butt (2014) 'Feelings as Agents of Selection: Putting Charles Darwin Back into (Extended Neo-) Darwinism', *Biological Journal of the Linnean Society* 112, 332–353.

Palmer, Clare (2003) 'Animals in Christian Ethics: Developing a Relational Approach', *Ecotheology* 7:2, 163–185.

Parsons, Glenn (2008) *Aesthetics and Nature*, London: Continuum.

Passmore, John (1974) *Man's Responsibility for Nature: Ecological Problems and Western Traditions*, New York: Charles Scribner's Sons.

Phemister, Pauline (1991) 'Leibniz, Freedom of Will and Rationality', *Studia Leibnitiana* 23:1, 25–39.

Phemister, Pauline (1999) 'Leibniz and the Elements of Compound Bodies', *British Journal for the History of Philosophy* 7:1, 57–78.

Phemister, Pauline (2003) 'Exploring Leibniz's Kingdoms: A Philosophical Analysis of Nature and Grace', *Ecotheology* 7:2, 126–145.

Phemister, Pauline (2004) '"All the Time and Everywhere Everything's the Same as Here": The Principle of Uniformity in the Correspondence between Leibniz and Lady Masham', in Paul Lodge (ed.), *Leibniz and His Correspondents*, Cambridge: Cambridge University Press, 193–213.

Phemister, Pauline (2005) *Leibniz and the Natural World: Activity, Passivity and Corporeal Substances in Leibniz's Philosophy*, Dordrecht: Springer.

Phemister, Pauline (2006) 'Progress and Perfection of World and Individual in Leibniz's Philosophy, 1694–1697', in Herbert Breger, Jürgen Herbst and Sven Erdner (eds), *Einheit in der Veilheit*, Proceedings of the VIII Internationaler Leibniz-Kongress, Hannover, 24–29 July 2006, 3 vols, Hannover: Gottfried-Wilhelm-Leibniz-Gesellschaft, vol. 2, 805–812.

Phemister, Pauline (2007) 'God's Freedom to Create', *Revue Roumaine de Philosophie* 51:1–2, 3–19.

Phemister, Pauline (2011) 'Monads and Machines', in Justin Erik Halldór Smith and Ohad Nachtomy (eds), *Machines of Nature and Corporeal Substances in Leibniz*, Dordrecht: Springer, 39–60.

Phemister, Pauline (2012) 'Relational Space and Places of Value', in Emily Brady and Pauline Phemister (eds), *Human–Environment Relations: Transformative Values in Theory and Practice*, Dordrecht: Springer, 17–30. Earlier revision in *Logical Analysis and History of Philosophy* 14 (2011), 89–106.

Phemister, Pauline (2015) 'The Souls of Seeds', in Adrian Nita (ed.), *Leibniz's Metaphysics and Adoption of Substantial Forms*, Dordrecht: Springer, 125–141.

Phemister, Pauline (2016) 'Leibnizian Pluralism and Bradleian Monism: A Question of Relations', *Studia Leibnitiana*, sonderhefte 45, 61–79.

Phemister, Pauline and Lloyd Strickland (2015) 'Leibniz's Monadological Positive Aesthetics', *British Journal for the History of Philosophy* 23:6, 1214–1234.

Plumwood, Val (2002) *Environmental Culture: The Ecological Crisis of Reason*, London: Routledge.

Rateau, Paul (2008) *La Question du mal chez Leibniz: Fondements et élaboration de la Théodicée*, Paris: Honoré Champion.

Ravaisson, Félix (1885 [1867]) *La philosophie en France au XIXe siècle*, 2nd edn, Paris: Librairie Hachette.

Renouvier, Charles (1912 [1864]) *Essais de critique générale: Troisième essai; Les principes de la nature*, 3rd edn, Paris: Armand Colin.

Renouvier, Charles and Louis Prat (1899) *La nouvelle monadologie*, Paris: Armand Colin.

Rescher, Nicholas (1979) *Leibniz: An Introduction to His Philosophy*, Oxford: Basil Blackwell.

Rice, Lee C. (1996) 'Spinoza's Relativistic Aesthetics', *Tijdschrift voor Filosofie* 58:3, 476–489.

Riley, Patrick (1996) *Leibniz' Universal Jurisprudence: Justice as the Charity of the Wise*, Cambridge, MA: Harvard University Press.

Rossi, Paolo (1984) *The Dark Abyss of Time: The History of the Earth and the History of Nations from Hooke to Vico*, trans. Lydia G. Cochrane, Chicago: University of Chicago Press.

Ruin, Hans (1998) 'Heidegger and Leibniz on Sufficient Reason', *Studia Leibnitiana* 30:1, 49–67.

Russell, Bertrand (1992 [1900]) *A Critical Exposition of the Philosophy of Leibniz*, London: Routledge.

Rutherford, Donald (1995) *Leibniz and the Rational Order of Nature*, Cambridge: Cambridge University Press.

Rutherford, Donald (2004) 'Idealism Declined: Leibniz and Christian Wolff', in Paul Lodge (ed.), *Leibniz and His Correspondents*, Cambridge: Cambridge University Press, 214–237.

Rutherford, Kenneth M. D., R. D. Donald, Ramona D. Donald, Alistair B. Lawrence and Françoise Wemelsfelder (2012) 'Qualitative Behavioural Assessment of Emotionality in Pigs', *Applied Animal Behaviour Science* 139:3, 218–224.

Saito, Yuriko (2007) *Everyday Aesthetics*, Oxford: Oxford University Press.

Schelling, Friedrich Wilhelm Joseph von (1856–61) *Friedrich Wilhelm Joseph von Schelling's Sämmtliche Werke*, ed. Karl Friedrich Schelling, 14 vols, Stuttgart and Augsburg: J. G. Cotta.

Schelling, Friedrich Wilhelm Joseph von (2004 [1799]) *First Outline of a System of the Philosophy of Nature*, trans. Keith R. Peterson, Albany, NY: State University of New York Press.

Schönfeld, Martin (2000) *The Philosophy of the Young Kant: The Precritical Project*, Oxford: Oxford University Press.

Serres, Michel (1968) *Le système de Leibniz et ses modèles mathématiques*, Paris: Presses Universitaires de France.

Sessions, George (1977) 'Spinoza and Jeffers on Man in Nature', *Inquiry: An Interdisciplinary Journal of Philosophy* 20:1, 481–528.

Simmons, Alison (2001) 'Changing the Cartesian Mind: Leibniz on Sensation, Representation and Consciousness', *Philosophical Review* 110:1, 31–75.

Simons, Peter (2015) 'Bolzano's Monadology', *British Journal for the History of Philosophy* 23:6, 1074–1084.

Smerlak, Matteo and Carlo Rovelli (2007) 'Relational EPR', *Foundations of Physics* 37:3, 427–445.

Smith, Justin E. H. (2007) 'The Body-Machine in Leibniz's Early Physiological and Medical Writings: A Selection of Texts with Commentary', *Leibniz Review* 17, 141–179.

Smith, Justin E. H. (2011) *Divine Machines: Leibniz and the Sciences of Life*, Princeton: Princeton University Press.

Smith, Justin E. H. and Pauline Phemister (2007) 'Leibniz and the Cambridge Platonists in the Debate over Plastic Natures', in Pauline Phemister and Stuart Brown (eds), *Leibniz and the English-Speaking World*, Dordrecht: Springer, 95–110.

Smith, Justin E. H. and Ohad Nachtomy (eds) (2011) *Machines of Nature and Corporeal Substances in Leibniz*, Dordrecht: Springer.

Spinoza, Benedictus de (1925) *Opera: Im Auftrag der Heidelberger Akademie der Wissenschaften*, ed. Carl Gebhardt, 4 vols, Heidelberg: Carl Winters.

Spinoza, Benedictus de (1985) *The Collected Works of Spinoza*, ed. and trans. Edwin Curley, vol. 1, Princeton: Princeton University Press.

Spinoza, Benedictus de (1995) *Spinoza: The Letters*, ed. and trans. Samuel Shirley, Indianapolis: Hackett Publishing Co.

Sprigge, Timothy L. S. (1971) 'Final Causes', *Proceedings of the Aristotelian Society, Supplementary Volume* 45, 149–170.

Sprigge, Timothy L. S. (1980) 'The Importance of Subjectivity: An Inaugural Lecture', in *The Importance of Subjectivity: Selected Essays in Metaphysics and Ethics*, ed. Leemon B. McHenry, Oxford: Clarendon Press, 2010.

Sprigge, Timothy (1984) *The Vindication of Absolute Idealism*, Edinburgh: Edinburgh University Press.

Stan, Marius (2013) 'Kant's Third Law of Mechanics: The Long Shadow of Leibniz', *Studies in History and Philosophy of Science* 44:3, 493–504.

Strawson, Galen (2006) 'Realistic Monism: Why Physicalism Entails Panpsychism', in Anthony Freeman (ed.), *Consciousness and Its Place in Nature: Does Physicalism Entail Panpsychism*, Exeter: Imprint Academic, 3–31.

Strickland, Lloyd (2006) *Leibniz Reinterpreted*, London: Continuum.

Strickland, Lloyd (2009) 'Leibniz, the "Flower of Substance," and the Resurrection of the Same Body', *Philosophical Forum* 40:3, 391–410.

Strickland, Lloyd (2010) 'The Doctrine of "the Resurrection of the Same Body" in Early Modern Thought', *Religious Studies* 46, 163–183.

Swoyer, Chris (1995) 'Leibnizian Expression', *Journal of the History of Philosophy* 33:1, 65–99.

Taliaferro, Charles (2001) 'Early Modern Philosophy', in Dale Jamieson (ed.), *Companion to Environmental Philosophy*, Oxford: Blackwell, 130–145.

Taylor, Paul W. (1986) *Respect for Nature: A Theory of Environmental Ethics*, Princeton: Princeton University Press.

Tissandier, Alex (2014) *Affirming Divergence: Deleuze's reading of Leibniz*, PhD thesis, Warwick: University of Warwick Publications Service & WRAP.

Vailati, Ezio (1997) *Leibniz and Clarke: A Study of Their Correspondence*, Oxford: Oxford University Press.

Vermenen, Patrice (1987) 'Les aventures de la force active en France: Leibniz et Maine de Biran sur la route philosophique menant à l'éclectisme de Victor Cousin', *Exercises de la Patience* 8, 147–168.

Voltaire [François-Marie Arouet] (2009 [1759]) *Candide, or All for the Best*, ed. and trans. Eric Palmer, Peterborough, ON: Broadview Press.

Walford, David (ed. and trans.) (1992) *Immanuel Kant: Theoretical Philosophy, 1755–1770*, in collaboration with Ralf Meerbote, Cambridge: Cambridge University Press.

Ward, James (1903 [1899]) *Naturalism and Agnosticism: The Gifford Lectures Delivered before the University of Aberdeen in the Years 1896–1898*, 2 vols, 2nd edn, London: Adam and Charles Black.

Ward, James (1920 [1911]) *The Realm of Ends or Pluralism and Theism: The Gifford Lectures Delivered in the University of St Andrews in the Years 1907–1910*, 3rd edn, Cambridge: Cambridge University Press.

Ward Thompson, Catharine and Peter Aspinall (2011) 'Natural Environments and Their Impact on Activity, Health and Quality of Life', *Applied Psychology: Health and Well-Being* 3:3, 230–260.

Ward Thompson, Catharine, Jenny Roe, Peter Aspinall, Richard Mitchell, Angela Clow and David Miller (2012) 'More Green Space Is Linked to Less Stress in Deprived Communities: Evidence from Salivary Cortisol Patterns', *Landscape and Urban Planning* 105, 221–229.

Watkins, Eric (1995) 'The Development of Physical Influx in Early Eighteenth-Century Germany: Gottsched, Knutzen, and Crusius', *Review of Metaphysics* 49:2, 295–339.

Watkins, Eric (1998) 'From Pre-established Harmony to Physical Influx: Leibniz's Reception in Eighteenth Century Germany', in Daniel Garber and Roger Ariew (eds), *Leibniz and the Sciences*, special issue, *Perspectives on Science* 6:1–2, 136–203.

Watkins, Eric (ed.) (2001) *Kant and the Exact Sciences*, Oxford: Oxford University Press.

Watkins, Eric (2006) 'On the Necessity and Nature of Simples: Leibniz, Wolff, Baumgarten, and the Pre-Critical Kant', *Oxford Studies in Early Modern Philosophy* 3, 261–314.

Watkins, Eric (ed. and trans.) (2012) *Immanuel Kant: Natural Science*, Cambridge: Cambridge University Press.

Wee, Cecilia (2001) 'Cartesian Environmental Ethics', *Environmental Ethics* 23:3, 275–286.

Whitehead, Alfred North (1960 [1929]) *Process and Reality: An Essay in Cosmology*, New York: Harper & Row.

Wilson, Catherine (1989) *Leibniz's Metaphysics: A Historical and Comparative Study*, Princeton: Princeton University Press.

Wilson, Catherine (1995a) *The Invisible World: Early Modern Philosophy and the Invention of the Microscope*, Princeton: Princeton University Press.

Wilson, Catherine (1995b) 'The Reception of Leibniz in the Eighteenth Century', in Nicholas Jolley (ed.), *The Cambridge Companion to Leibniz*, Cambridge: Cambridge University Press, 442–474.

Wilson, Catherine (2010) 'Leibniz's Reputation in the Eighteenth Century: Kant and Herder', in Jill Kraye, John Rogers and Tom Sorell (eds), *Insiders and Outsiders in Seventeenth-Century Philosophy*, London: Routledge, 294–308.

Winters, Edward (2007) *Aesthetics and Architecture*, London: Continuum.

Wolff, Christian (1962) *Christiani Wolffii Philosophia prima sive ontologia*, ed. Johannes École, *Gesammelte Werke*, vol. 3, pt 2, Hildesheim: Olms.

Wolff, Christian (1964) *Christiani Wolffii Cosmologia generalis*, ed. Johannes École, *Gesammelte Werke*, vol. 4, pt 2, Hildesheim: Olms.

Wolfson, Harry Austryn (1983 [1934]) *The Philosophy of Spinoza: Unfolding the Latent Processes of His Reasoning*, 2 vols (combined), Cambridge, MA: Harvard University Press.

Woolhouse, Roger S. (1993) *Descartes, Spinoza, Leibniz: The Concept of Substance in Seventeenth Century Metaphysics*, London: Routledge.

Yolton, John W. (1984) *Thinking Matter: Materialism in Eighteenth-Century Britain*, Oxford: Blackwell.

Yolton, John W. (1991) *Locke and French Materialism*, Oxford: Clarendon Press.

Index